LECTURES
ON THE WHOLE OF
ANATOMY

THE ROLLS PARK PORTRAIT OF WILLIAM HARVEY (*circa* 1622)

WILLIAM HARVEY

Lectures On the Whole of Anatomy

AN ANNOTATED TRANSLATION OF
Prelectiones Anatomiae Universalis
BY
C. D. O'MALLEY
F. N. L. POYNTER
K. F. RUSSELL

UNIVERSITY OF CALIFORNIA PRESS

BERKELEY AND LOS ANGELES : 1961

UNIVERSITY OF CALIFORNIA PRESS
BERKELEY AND LOS ANGELES

CAMBRIDGE UNIVERSITY PRESS
LONDON, ENGLAND

© 1961
BY THE REGENTS OF THE UNIVERSITY OF CALIFORNIA

LIBRARY OF CONGRESS CATALOG CARD NUMBER: 61–16879

DESIGNED BY ADRIAN WILSON

ACKNOWLEDGMENTS

IT IS OUR pleasant task to express our thanks to the following publishers. To Blackwell Scientific Publications Limited for permission to quote from Harvey, *The Movement of the Heart and Blood in Animals, an Anatomical Essay*, translated by Kenneth J. Franklin (1957). To the Clarendon Press for permission to quote from *The Works of Aristotle translated into English*, Volumes II (1930), III (1923), IV (1910), V (1911). The lines from Athenaeus, *The Deipnosophists*, Vol. IV, translated by C. B. Gulick (1930), *Scriptores Historiae Augustae*, Vol. III, translated by D. Magie (1932), Celsus, *De Medicina*, Vol. III, translated by W. G. Spencer (1953), and Galen, *On the Natural Faculties*, translated by A. J. Brock (1928), are reprinted by permission of Harvard University Press and The Loeb Classical Library.

We are grateful to the Royal College of Physicians of London for permission to employ one of the Baldwin Hamey manuscripts, and to the British Museum for permission to use the manuscript of Harvey's *Prelectiones* and to reproduce several pages from it. The present work was greatly aided by a grant from the National Science Foundation to C. D. O'M., and finally, and far from least, the appearance of the volume has been greatly enhanced through the inclusion of the sole authentic portrait of Harvey during the period when he was compiling the *Prelectiones*; this was made possible through the kind permission and assistance of its owner, Myron Prinzmetal, M.D., of Beverly Hills, California.

CONTENTS

INTRODUCTION   *1*

ABBREVIATED TITLES OF WORKS CITED FREQUENTLY   *20*

MAJOR DIVISIONS OF HARVEY'S TEXT

Definitions of anatomy. The parts of the body, their
distinctions and functions   1v   *22*

Procedures and purposes of anatomy   4r   *27*

Divisions of the human body, their relative proportions
and uses. The three venters of the body. Skin, hair, fat
and underlying membrane   7r   *31*

The lower venter (abdomen): investing muscles, peritoneum and contained parts, omentum   14v   *49*

The organs of digestion: stomach, intestines, mesentery, gall bladder, pancreas, kidneys, ureters, bladder 18v   *59*

The digestive process   33r   *89*

The genito-urinary system: kidneys, ureters, bladder,
genitalia   43r   *113*

The middle venter (thorax). The chest and its parts:
neck and its contents, ribs and sternum, diaphragm and
lungs   61r   *150*

The cardiovascular system. The motion of the heart
and pulse. The circulation of the blood. The significance of respiration   72r   *171*

The upper venter (head): shape, anatomy, and purpose   87v   *206*

The nervous system: brain, cerebellum, cranial nerves 89v   *210*

INDEX OF NAMES   *237*

ILLUSTRATIONS

THE ROLLS PARK PORTRAIT OF
WILLIAM HARVEY   *Frontispiece*

THE ROYAL COLLEGE OF PHYSICIANS AT
AMEN-CORNER   2

THE TITLE PAGE OF HARVEY'S NOTEBOOK   8

PAGE 5ʳ OF THE NOTEBOOK   29

PAGE 80ᵛ OF THE NOTEBOOK, SHOWING HARVEY'S
FAMOUS STATEMENT ON THE CIRCULATION
OF THE BLOOD   193

To the memory of
John Farquhar Fulton
physiologist, medical historian, and bibliophile

INTRODUCTION

ANATOMICAL TEACHING in Britain commenced officially in Edinburgh in 1505 when the magistrates of that city granted a Seal of Cause to the Guild of Surgeons and Barbers. This was confirmed by James IV in the following year.

In London the United Company of Barber-Surgeons was granted a Charter in 1540 by Henry VIII which enabled the Company to obtain the bodies of felons for dissection. There is ample evidence to suggest that from this date the barber-surgeons of London carried out anatomical teaching within their Company.

Anatomy was then in the hands of the surgeons, and it was some time before the physicians interested themselves in the subject. In this curious situation the more highly educated physicians looked down upon their lowly colleagues and, indeed, made it their business to harass them in their practice, particularly when the barber-surgeons dabbled in physic. The physicians, however, could not always ignore anatomy and, although the documentary evidence is scanty, it would seem that the College of Physicians entered the anatomical field in 1565 when Elizabeth I granted them the right of claiming the bodies of felons for dissection. As Munk points out, the first mention of anatomical lectures for physicians is to be found in the College Annals for 1569–1570. These were to be delivered in turn by the Fellows, and severe penalties were laid down for their default from giving the lecture.[1]

[1] William Munk, *The Roll of the College of Physicians of London* (London, 1878), I, 69.

1

THE ROYAL COLLEGE OF PHYSICIANS AT AMEN-CORNER, WHERE
HARVEY DELIVERED HIS LUMLEIAN LECTURES
(A NINETEENTH-CENTURY RECONSTRUCTION BY T. SPENCER)

2

The lectures were open to Fellows, Candidates, and Licentiates, and were given each year with varying regularity until the great fire of 1666 destroyed the College house in Amen-corner. Some time after this they were discontinued, being merged with or superseded by the Gulstonian lectures. They were independent lectures, quite unrelated to the Lumleian benefaction with which we are now more concerned.

In 1581 Richard Caldwell, a Fellow, and Lord Lumley obtained the queen's permission to found a surgical lecture at the College with an endowment of 40 pounds per year, and the College built an appropriate room for "the better celebration of this most solemn lecture."[2] The indenture establishing the lecture was received by the College on August 3, 1582. The best account of its foundation and scope is that given by Holinshed (1587):

> This yeare, 1582, was there instituted and first founded a publike Lecture or Lesson in Surgerie, to begin to be read in the College of Physicians, in London, in Anno 1584, the sixt day of Maie, against that time new reedified in a part of the house that doctor Linacre gaue by testament to them, by John Lumleie, Lord Lumleie, and Richard Caldwell, doctor in Physicke, to the honour of God, the common profit of hir Maiesties subiects, and good fame, with increase of estimation and credit, of all the surgians of this realme. The reader whereof to be a doctor of physicke, and of good practise and knowledge, and to haue an honest stipend no lesse than those of the vniversities erected by king Henrie the eight, namelie, of law, diuinitie, and physicke, and lands assured to the said college for the maintenance of the publike lesson; whereunto such statutes be annexed as be for the great commoditie of those which shall give and incline themselves to be diligent hearers for the obteining of knowledge in surgerie, as whether he be learned or vnlearned that shall become an auditor or hearer of the lecture, he may find himselfe not to repent the time so imploied. First, twise a weeke through out the yeare, to wit, on wednesdaies and fridaies, at ten of the clocke till eleuen, shall the reader read three-quarters of an houre in Latine, and the other quarter in English, wherein that shall be plainlie declared for those that vnderstand not Latine what was said in Latine. And the first yeare to read

² *Ibid.*, III, 352.

3

Horatius Morus tables, an epitome of briefe handling of all the whole arte of surgerie, that is, of swellings or apostems, wounds, ulcers, bonesetting, and healing of bones broken, termed commonlie fractions, and to read Oribasius of knots, and Galen of bands, such workes as haue beene long hid, and are scarselie now a daies among the learned knowen, and yet (as the anatomies) to the first enterers in surgerie and nouices in physicke; but amongst the ancient writers and Grecians well knowne. At the end of the yeare, in winter, to dissect openlie in the reading place, all the bodie of Man, especiallie the inward parts, for fiue daies togither, as well before as after dinner; if the bodies may so last without annoie.

The second yeare to read Tagaultius institutions of surgerie, and onelie of swellings or apostems, in the winter to dissecte the trunk onelie of the bodie, namelie, from the head to the lowest part where the members are, and to handle the muscles especiallie. The thirde yeare to read of wounds onelie of Tagaultius, and in winter to make publike dissection of the head onelie. The fourth yeare to read of vlcers onlie the same author, and to anatomize or dissect a leg and an arme for the knowledge of muscles, sinewes, arteries, veines, gristles, ligaments, and tendons. The fift year to read the sixt booke of Paulus Aegineta , and in winter to make anatomie of a skeleton, and therewithall to show and declare the use of certeine instruments, as Scamnum Hippocratis, and other instruments for setting in of bones. The sixt year to read Holerius of the matter of surgerie, as of medicines for surgians to use. And the seventh yeare to begin again and continue still. A godlie and charitable creation doubtlesse, such as was the more needfull, as hitherto hath beene the wante and lacke so hurtfull; sith that onelie in ech vniversities by the foundation of the ordinarie and publike lessons, there is one of physicke, but none of surgerie, and this onelie of surgerie and not of physicke, I mean so as physicke is now taken separatelie from surgerie and that part which onelie vseth the hand as it is sorted by the apothecarie. So that now England may rejoise for those happie benefactors and singular welwillers to their countrie who furnisheth hir so in all respects, that now she may as compare for the knowledge of physicke, so by means to come to it, with France, Italie, and Spaine, and in no case behind them, but for a lecture in simples, which God at his pleasure may procure, in mouing some hereafter in like motion and instinct to be as carefull and beneficiall as these were to the helpe and furtherance of their countrie. At the publication of this foundation, which was celebrated with a goodlie assemblie of doctors, collegiats, and licentiats, as also some masters of surgerie, with other students, some whereof had beene academicall; doctor Caldwell so aged that his number of yeeres with his white head adding double reuerence to his person (whereof I may well saie no lesse than is left written of a doctor of the same facultie verie famous while he liued

> Conspicienda aetas, sed et ars prouectior annis,
> Famaque paeonio non renuenda choro).

Euene he, notwithstanding his age and impotencie, made an oration in Latin to the auditorie, the same by occasion of his manifold debilities unfinished at the direction speciallie of the president who (after a few words, shortlie and sweetlie, vttered) gaue occasion and opportunitie to D. Forster, then and yet the appointed Lecturer, to deliuer his matter, which he discharged in such methodicall manner, that ech one present indued with judgment conceiued such hope of the doctor, touching the performance of all actions incident unto him by that place, as some of them continued his auditors in all weathers and still hold out; whose diligence he requiteth with the imparting of further knowledge than the said publike lecture doth afford. When the assemblie was dissolued, and the founder accompanied home, diligent care was taken for the due preferring of this established exercise: insomuch that D. Caldwell and D. Forster, to furnish the auditors with such bookes as he was to read, caused to be printed the epitome of Horatius Moras, first in Latine: then in English, which was translated by the said doctor Caldwell. But before it was half perfected, the good old doctor fell sicke, and as a candle goeth out of it selfe, or a ripe apple falling from the tree, so departed he out of this world, at the doctors commons, where his vsuall lodging was; and was verie worshipfullie buried.[1]

Richard Forster was the first Lumleian lecturer. Although the appointment was for life he resigned in December, 1602, having delivered the whole course of lectures three times—not always, it seems, to so numerous an audience as that which Holinshed tells us attended his opening lecture on May 6, 1584, for even in the first year a complaint of poor attendance was recorded in the College *Annals*. Forster was succeeded by William Dunne, who died in May, 1607, and then by Thomas Davies, whose death in

[1] R. Holinshed, *The First and Second* [and third] *Volumes of Chronicles . . . now newlie augmented and continued . . . to the yeare 1586 by John Hooker*, III (London, 1587), p. 1349.

August, 1615, when Forster himself was President of the College, led to Harvey's appointment. Harvey was to carry out the duties of this office for more than 40 years, for he did not resign until July, 1656.

Some of the personal notes that Harvey made for his first course of lectures survived the destruction of many of his papers during the Civil War and eventually came into the possession of Sir Hans Sloane, with whose collection they passed to the British Museum, where they may be seen and consulted. In 1886 the Royal College of Physicians published a facsimile reproduction together with a transcription of these *Prelectiones Anatomiae Universalis*, [4] as Harvey himself called them. Written in a difficult hand, presumably for no eyes other than his own, they are mainly in Latin, with occasional words and phrases in English. In the following pages they are presented for the first time in an entirely English version, with appropriate annotations and identifications of the numerous references to Harvey's literary sources.

In the introduction to the facsimile edition it is stated that "The object of the publication was to preserve and make public the original notes of the lectures in which Harvey, in 1616, set forth for the first time his discovery of the circulation."[5]

The circumstances attending the production of these lecture notes have never been discussed, but the more closely they are investigated the clearer does it become that many features of them which have been taken for granted are still open to question. They are certainly notes that Harvey prepared for his Lumleian lectures, and, judging by their scope, by the research into the literature which is revealed in the citations, and by the personal observations briefly referred to, Harvey must have spent much time in compiling them. It has always been assumed that they were the rough notes for anatomical lectures delivered at a dissection carried out on April 16, 17, and 18 in the year 1616.

[4] *Prelectiones Anatomiae Universalis by William Harvey. Edited with an autotype reproduction of the original by a committee of The Royal College of Physicians of London*. London: Churchill, 1886. The transcription was made by Edward Scott of the Manuscript Department of the British Museum.
[5] *Ibid.*, p. v.

Highly condensed as they are, we are forced to wonder how Harvey conveyed so much information, accompanying it with actual demonstration of dissected parts, in the brief time available. It is possible, of course, that much was omitted in the actual lectures, for the notes must have been written with more than the single dissection in mind.

Let us consider the order of events. Harvey was appointed Lumleian lecturer in August, 1615, but presumably he would have been given a little time to make his preparations. If he followed the schedule laid down in the terms of the foundation he would have spent his first year of office by lecturing on Wednesdays and Fridays, from 10 a.m. to 11 a.m., on the whole art of surgery, with Horatius Morus, Galen, and Oribasius as the prescribed texts. "At the end of the year, in winter," he was "to dissect . . . all the body of man, especially the inward parts, for five days together, if the bodies may so last without annoie. . . . " It seems Harvey decided in advance that three days was as long as he wished to spend over a cadaver,[6] so why, if the generally accepted story is correct, should he choose to depart from the terms of the foundation and, little more than six months after his appointment, embark on a dissection in the second half of April, which can often bring warm days in London.

But is this in fact what happened? The title page has at least one apparent error that seems to have escaped notice: a fact implying that this page merits closer study. Harvey was actually 38 years of age on April 1, 1616, and on the 16th of that month was in his 39th year, not his 37th. The College transcript correctly indicates that the year "1616" is crossed through after the word "Aprili" and the numerals "16," "17," and" 18" are written on the following line, but it omits the letters written beneath those numerals which are clearly visible in the facsimile, as follows: 16 17 18. Before offering any suggestion as to the
    d   J   f
significance of these letters, we should recall that his office required Harvey to dissect and lecture on the trunk and all

---

[6] *Ibid.*, f. 4^r.

THE TITLE PAGE OF HARVEY'S NOTEBOOK

the organs contained therein at the end of his second year (1617), on the head only at the end of the third year (1618), and on the leg and arm, with special reference to the muscles, sinews, and ligaments, at the end of the fourth year (1619). In the winter of the fifth year he was "to make an anatomy of a skeleton" and demonstrate the bones and the setting of dislocations and fractures; the last year of the course required no dissection. In 1622 he began the course again and in November of that year we find that Theodore Goulston was excused by the President from carrying out a dissection because Harvey ex officio was doing this himself and lecturing on the whole of anatomy.

Examination of the notes that follow will show that by the standards of his day Harvey had ample material, not only for the general survey of anatomy to be given at the end of the first year, but also for the more detailed study of the trunk and the head to be given at the end of the second and third years respectively. An interpretation of the information given on Harvey's own title page which may be nearer the truth than the current oversimplification, would suggest that Harvey took up his duties as Lumleian lecturer in April, 1616,[7] and he began then to make the notes that are translated in this volume—the more detailed notes for the thorax and the head being prepared for the anatomical lectures of 1617 and 1618. When these were added to the growing pile of notes, he turned back to the title page, crossed through the year "1616" and added in the next line the figures 16, 17, and 18. The supply of anatomical material was strictly limited and Harvey could not have known with certainty the precise date when a cadaver would become available. After the dissections had been performed he added the letters below the figures, which could well mean "D[ecember 16]16," "J[anuary 16] 17" (i.e., 1618, for the year ended on March 24), and "F[ebruary 16]18" (i.e., 1619). These dates are strictly in accordance with the terms of the Lumleian foundation and are supported by an important statement by Harvey himself to

[7] Munk, *op. cit.*, Vol. I, p. 126, where Munk, the College annalist, states quite definitely that "Harvey commenced his lecture in April, 1616."

which we shall refer in a moment. Immediately following this series of notes should come the 68 pages of similar notes on the anatomy of the leg and arm which are found at the beginning of a further volume, also a Sloane MS in the British Museum, containing the treatise *De motu locali animalium* (recently translated by Dr. Gweneth Whitteridge). We thus have the dissection notes for the whole course, with the exception of notes on the skeleton, which Harvey may have thought unnecessary. Having compiled his notes, Harvey used them again and again, as is the custom with many lecturers, adding others in the margins or on blank versos or even inserting odd leaves as occasion required. The several conflicting series of numerals in the notes, suggesting differing emphases, also support this view. The numbering of the leaves is certainly not Harvey's and there is no evidence whatever that folio 80ᵛ, for example, which contains his most precise statement of the circulation, was not written later—even an appreciable amount of time later—than folio 81. The earliest date at which some of the notes could have been written can be established within fairly narrow limits. There is a pathological note on folio 26ᵛ which can refer only to an autopsy on his own father, who died on June 12, 1623; a similar note on folio 54ᵛ alludes to Lord Chichester, the Lord Deputy of Ireland, who died on February 19, 1625; a page reference on folio 49ᵛ shows that Harvey consulted an edition of Riolan's *Anthropographia et Osteologia* that was published in Paris in 1626.

It is clear that by no means all the notes were written before April, 1616, nor can they represent the substance of lectures delivered only at that time. If our interpretation of the evidence is correct, the earliest date on which the brief account of the circulation noted on folio 80ᵛ could have been presented in a lecture is January, 1618, when Harvey carried out the second year's anatomical dissection. That this is highly probable and that this particular note was not written for a similar lecture when it came round again in 1623 is borne out by Harvey's own statement. In the dedicatory epistle to the President and Fellows of the College which prefaces *De Motu Cordis* in 1628, Harvey says: "On several earlier occasions in my anatomical lectures I revealed my new concept of the heart's movement and function and of the blood's passage round the body. Having now, however, for more than nine years confirmed it in your presence . . . I have . . . published it for all to see."[8] "More than nine years" does not mean twelve, which is the figure Harvey could have written if he had indeed first demonstrated the circulation in April, 1616, and it is most unlikely that he meant anything but precisely what he wrote; that is, more than nine but not as long as ten years: a statement that accords completely with January, 1618, as the date of that memorable occasion.

Having removed what we believe to be some common misconceptions concerning this manuscript, we must consider its relative importance in the Harveian canon. *De Motu Cordis* is universally acknowledged to be the first brilliant demonstration of modern scientific method. These lecture notes, in which we see Harvey, the confirmed Aristotelian, gradually reaching out toward the new method and weighing the results of his own observation and experiment against the opinions of others, give us a fascinating prelude to that great work. Its range, especially in comparative anatomy and pathology, is extremely wide and can be matched only in the work of John Hunter 150 years later. In a recent study of "Harvey's animals," the late F. J. Cole recorded no less than 128 types of animal life that are mentioned in these notes.[9] It is true that many of these occur in references to Aristotle and other writers, but an impressive number represent personal observations and dissections made by Harvey himself of fish, fowl, serpents, and quadrupeds, among these the rats, mice, and dogs familiar in the modern research laboratory. Equally impressive is the scope of his reading in both ancient and contemporary authors. The numerous citations throughout the notes provide a useful indication of the content of Harvey's library. His supreme authority was

[8] *De Motu Cordis*, p. 5.
[9] F. J. Cole, "Harvey's Animals," *J. Hist. Med.* (1957), 12: 106–113.

Aristotle, who is quoted on nearly every page. Galen too is often called upon, and so reluctant was Harvey to refute some of Galen's errors in human anatomy that he suggests men have changed in 2,000 years. The same charity was accorded by Harvey to his contemporaries: Caspar Bauhin was "a rare industrious man," and even when disagreeing with Du Laurens he adds, "peace to that great man."

What Harvey considered to be his own original observations are usually prefaced with his monogram **WH**, as are the notes on the circulation on folio 80ᵛ, but there are many other noteworthy passages, such as his early account of a cirrhotic liver when he tells of a case of ascites which at autopsy showed the liver grey, retracted, and hard, or his account of the processus vaginalis testis which must have been based on dissection of the foetus. When discussing the suprarenal gland he mentions that "often when the kidneys have been removed it adheres to the diaphragm." This fact is one that could have been noted only in dissection, for the suprarenal is embedded in the fat above the kidney, which is probably why these glands were not mentioned by Vesalius, although Eustachius had fully described them.

It is interesting to find here the earliest reference to the embryological researches that were not to be published until 1651, in which he says that in the developing chick the heart appears before the liver: "I have seen as Dr. Argent will be my witness, all things perfected with the liver unformed. The heart [is] formed with the auricles [when] the liver is still a rude and confused mass. The heart very white, the auricles reddish and filled with blood."

These notes also reveal Harvey as the "complete doctor" at a time when knowledge and the search for it were not yet compartmentalized. His *De Motu Cordis* would have crowned with glory the career of any full-time research worker, and its character tends to make us overlook that Harvey was also a busy practitioner and a pathologist as well as experimental biologist, anatomist, and physiologist. The clinical approach is evident throughout his teaching and he cites many of his own cases drawn from his practice in both London and the home counties, especially his native county of Kent. Some of his patients are mentioned by name and their condition used to illustrate some anatomical point: Sir William Rigdon, with his cholelithiasis and jaundice, must have worried or impressed Harvey; Lady Croft, "always complaining"; Mrs. Young, with her spleen so enlarged as to counterfeit pregnancy; John Bracy with his enlarged liver "as big as an ox liver"; his own sister, whose enlarged spleen (probably from malaria) weighed five pounds; Joan Johnson dead from a malignant fever and hematemesis; Mrs. Chirn with a pyronephrosis; Sir Thomas Hardes with urethral stricture; Sir Robert Wroth with renal calculi. Among those unnamed are the man behind Covent Garden with the huge hydrocele and the man with a wound in the chest "that blew out a light." Despite Aubrey's derogatory remarks about Harvey as a practitioner, there is evidence that his own colleagues entrusted their families to his care: he treated Dr. Argent's daughter for severe headaches, and Dr. Argent himself, as well as Theodore Goulston, may have been his patients, for he mentions the former's fistula and the latter's gallstones in his notes. It is not considered good form nowadays for a doctor to discuss publicly his "cures," but in Harvey's days it was commonplace. Thus, from his notes we gather that Harvey claimed to have cured cases of diabetes, dropsy, venereal diseases, prolapsus uteri, varicose hernia, inguinal hernia, hydrocele, and many other conditions, and that he even practiced surgery himself and had actually performed lithotomy.

In all discussions on the history of anatomy the restrictions on the supply of cadavers for dissection are always emphasized, but certain facts in these notes bring home to us that these were not the only sources of knowledge of human anatomy. Autopsies were apparently carried out regularly and, it seems, without any of the inhibitions imposed by sentiment or social prejudice. Harvey's own father and sister were among the cases whose histories were thus completed at autopsy, and there are many others that enabled Harvey to extend his pathological knowledge, as for instance of the tubercular adhesions of the lung and the realization that "one part of a lung hath served some a long time," and, similarly, the observation that man can live with a single kidney.

For the careful and sensitive reader these notes provide many hints of the manner in which Harvey enlivened his lectures with personal anecdotes and reminiscences. There are echoes of his life in Cambridge and Italy. Although we have found no other record of the event it seems that Harvey did attend at least one dissection at Cambridge. At Padua these opportunities were more frequent and included dissection of a Paduan courtesan and of a man struck by lightning. When discussing liver abscesses he recalls: "I observed these things in the hospital [i.e., St. Bartholomew's], as well as in the hospitals of Italy, with much nausea, loathing and foetor. I have forgotten many things." He notes that he had seen the spleen dissected in wolves, presumably in Italy, and when he mentions greyhounds, falcons, hawks, and fighting cocks, or compares the convolutions of the small intestine to an unset ruff, or illustrates the difference of gait by referring to the "executioner's walk," he reminds us of the age in which he lived. We are also reminded of the city in which he lectured and practiced by his comparison of the alimentary tract to a thoroughfare near St. Paul's, each part of which has separate names, and by his allusion to the royal aviary in St. James's Park. Occasionally there are fascinating glimpses of the more intimate side of Harvey, as when he discusses the prehensile feet of apes and then reveals that "I have learnt to grasp with my feet," or when he refers to the passage of indigestible objects through the body and adds, "gold ring—mine," apparently swallowed and recovered. It is also interesting to see how closely this childless man watched children playing football and other sports, and to note how even such incidents provided grist for his mill. Sometimes a few words are sufficient to recall familiar episodes in his later life. He notes that "if the body is constricted by cold [with] bare feet there is copious urine," and we are reminded by Aubrey that when he was nearing seventy and afflicted with gout he "would sit with his legs bare, though it were frost, on the leads of Cockaine House, put them into a pail of water till he was almost dead with cold," and how, when his thoughts kept him awake, he would "rise from his bed and walk about his chamber in his shirt till he was pretty cool and then return to his bed and sleep very comfortably."[10]

In his study *The Personality of William Harvey*, Sir Geoffrey Keynes remarks, "There is, indeed, little opportunity in Harvey's writings of judging whether he possessed a sense of humour at all, though it would be rash to assume from this that he did not possess one."[11] In these pages will be found several examples of a dry, sardonic humor which goes well with all that we know of Harvey's character. Discussing the calloused skin produced by continual friction or pressure, he writes, "spurred Hackneys, saints on their knees"—saints being a term applied to the overenthusiastic puritans of his day. Harvey was no Parliament man, but a royalist, and he compares the acid taste rising from the stomach to a motion sent from the lower to the upper house of Parliament. It may well be that this same satirical spirit infuses remarks that could be taken too seriously and cited as examples of vestigial medievalism in Harvey, as when he writes, "The neck for adornment, for all are adorned"—a remark that can be matched by his opinion of fashion in dress. "Fashion," he says, "is but a redundant covering, a fantastic arrangement," and "the best fashion to lep to run to doe anything stripp to ye scin." He has noticed the rouge clinging in the wrinkles on the faces of the Venetian courtesans, and reminds his audience that "the scin [is] not for paynting."

Interesting and important as these notes may be for the light they throw on the life and work and character of William Harvey, the closest attention will naturally be concentrated on the notes on the heart which foreshadow the little book that is one of the greatest classics of scientific literature. It is clear that when he came to write the notes, he had been making his experiments and observations on the heart and circulation for a long time. When

[10] J. Aubrey. *Brief Lives, set down . . . between the years 1669 and 1696*, ed. Andrew Clark. Oxford, 1898.
[11] (Cambridge, 1949), p. 14.

talking of the systole and diastole of the heart, on folio 77ᵛ, he writes, "Having observed [the motion of the heart] for whole hours at a time, I was unable to discern [these things] easily by sight or touch, wherefore I propose that you ought to observe and note." Certainly he had already come to definite conclusions about the circulation, for on folio 79ᵛ he states dogmatically: "The heart having been extended and contracted, just as by a kind of force it propels from the right [ventricle] into the lungs, from the left into the aorta; wherefore [occurs] the pulse of the arteries. . . . " Again, on folio 78ᵛ, he says, "Hence the pulse of the artery [is] not from an innate faculty of the valves, as according to Galen 13, but by the heart thrusting forth [as is indicated] by autopsy in the live and dead, by reason [and] by experiment with ligatures." His views on the origin of the heart beat are clearly set down on folio 77ᵛ: "The pulse begins at the auricles and progresses to the point, wherefore as if [there were] two wings WH. Nevertheless the heart beats when separated from the auricles and the auricles awaken the somnolent heart." These remarks could only follow prolonged observations on dissections of living animals. Finally, on folio 80ᵛ is the conclusive statement, " . . . wherefore the beat of the heart produces a perpetual circular motion of the blood."

The section on the heart, although short, is one of the best in his notebook; it is certainly the most important. The detailed anatomy of the heart is good and his views are clear-cut. He at once enters into argument with Galen when he states that the veins arise in the heart, and not in the liver as in the ancient teaching. He crosses swords with Aristotle in the description of the ventricles:

> There are two ventricles, right and left. WH I am astonished at Aristotle since he describes three so precisely. Not unless it is possible to exclude the left auricle from the ventricles in which matter Galen rightly blames him since the ventricles are the same in the horse and the sparrow. Therefore WH I doubt whether animals have undergone such great alteration in the passage of time. It is strange that an author so diligent and faithful should err.

Here Harvey talks with the dogmatism obtained by experience and dissection; here, too, one senses a certain testiness. The heart with three ventricles was also a medieval concept—so described by Mondino who borrowed it in turn from Avicenna—now neatly disposed of by Harvey.

It is from the fabric of these short notes on the heart that *De Motu Cordis* was to be built. From 1616 until after the publication of *De Motu Cordis* there is no contemporary account of the impact of Harvey's lectures on his audience.[12] However, in the Royal College of Physicians there is a manuscript of the anatomical lectures given by Baldwin Hamey (1600–1676), one of the great benefactors of the College, on January 22, 24, and 25, 1647/48.[13] The manuscript, in Hamey's hand, gives us an excellent picture not only of the author's learning but also of the full content of lectures of this type. He had attended Harvey's lectures and obviously profited from them. When discussing the movement of the blood, Hamey gives a discussion on the theories put forward by the ancients and his predecessors and then says:

> From the left ventricle, it is, as I sayd, expel'd by the arteries into the whole body; and there, till of late yeares, it was thought to rest, nor was there any further heed taken, or account given of it, than this; that it served for nutrition and augmentation, for generation of spirits and of sperme in their due times. But now by the conduct of our renowned Professor and Colleague Dr. Harvey, there is a way found to bringe the greatest part back againe, and yet no part of the foresaid worke left undone. So that now we truely know what . . . is meant by Circularis Disciplina, for it may be shew'd us in every body that hath a Heart.

Harvey's years of teaching and "ocular demonstrations" at the College had not been wasted.

As was mentioned above, Harvey's lecture notes were deposited in the British Museum as part of the manuscript collection of Sir Hans Sloane. In 1877, after attention had

[12] For an echo of the lectures in one of John Donne's sermons, see F. N. L. Poynter, "John Donne and William Harvey," *J. Hist. Med.* (1960), 15:233–246.

[13] B. Hamey. *Praelectiones anatomicae habitae in Collegio medico 22, 24, 25 Januarii, 1647/8 Balduino Hameo M.D. Authore.* Manuscript RCP 143, Royal College of Physicians, London.

been called to the significance of these notes, a proposal was made for their publication in a facsimile edition, and with the financial support of the Royal College of Physicians and of the Royal College of Surgeons a facsimile edition with printed transcription was published in 1886. A comparison of the facsimiles with the manuscript itself indicates that they were executed as faithfully as photography of that period permitted. A few marginal annotations in red ink are not visible in the facsimile, but they have become so faded in the course of time that they are scarcely legible in the manuscript. However, although the editorial committee carried verisimilitude to the extent of reproducing some blank pages from the manuscript, they failed to include any mention of the fact that there are three wholly blank and unused leaves in it, properly folios 42–44; or the fact that on two folios, 66 and 67, Harvey, about to write on the versos, turned these leaves top to bottom and so produced a margin on the left side of the page. In the facsimiles, as a consequence, these pages create the illusion of their having been bound in erroneous sequence.

Despite the well-known difficulties of Harvey's handwriting, and especially in notes written for himself, the transcription deserves commendation. Occasionally, however, ignorance of the subject of the notes led the transcriber into error, and sometimes in his perplexity he produced wholly meaningless combinations of letters. We have indicated such lapses and given the correct transliteration, or our conjecture if uncertain, together with the appropriate translation into English.

For the most part Harvey employed Latin for his notes, frequently ungrammatical and sometimes thereby introducing difficulties for translation. From time to time he slipped from Latin into English, and all such words in English have been retained (in italics), as well as his few words of French or Italian. One gains the impression that Harvey was not fully a master of the Latin language, and that as certain thoughts and ideas occurred to him he found it much easier and quicker to express them in the vernacular rather than to search for Latin equivalents. Of course the lectures as delivered would contain no indication of such deviation from the classical tongue. The few Greek terms employed have been transliterated into roman letters.

Since Harvey's notes were written solely for his own use they are frequently so abbreviated as to be cryptic, and mere translation of them would have been meaningless. From time to time, therefore, it has been necessary to extend his remarks, but all such extensions have been indicated by inclusion within brackets. On the other hand, such extension has not been employed if it was felt that the reader could supply the necessary expansion of thought or the missing link between phrases. Where explanation of meaning was too extensive to be placed within the text it has been relegated to the notes. The general principle followed has been that of presenting Harvey's own words obscured as little as possible by interpretive guesses, and in such puzzles as remain it is hoped that the readers will be able to supply further emendations. Throughout his notes Harvey employs three symbols: **WH**, Δ, and **X**. The first, as his monogram, appears to indicate that he personally, as contrasted to others, has observed or can vouchsafe the particular phenomenon. The Δ, as it has been asserted, represents something demonstrated, and the **X** merely calls attention to, or is the equivalent of, cf. In practice, however, his use of these symbols is not always limited so precisely, and consequently they have been retained as he employed them.

It is in these notes more than in any other of his writings that we discover those authors most consulted by Harvey, and in some instances reference to a page number permits us to determine which edition of a work he consulted. However, it should be remarked that since the notes were composed over a period of years, reference to a particular edition of a work means only that at some time in the course of their composition did Harvey consult that particular edition. Thus, for example, at one time he employed the *Observationes Anatomicae* of Fallopius in the first edition of 1561 and at another in the second edition of 1562. Where Harvey's use of a specific edition of a work could be determined, it has been mentioned in the notes.

19

Abbreviated titles of works cited frequently

Bauhin Caspar Bauhin. *Theatrum Anatomicum*. Basel, 1605.
Colombo Realdo Colombo. *De Re Anatomica*. Frankfurt, 1590.
Crooke Helkiah Crooke. *Description of the Body of Man*. London, 1631.
Daremberg *Oeuvres Anatomiques, Physiologiques et Médicales de Galien*, trans. Ch. Daremberg. Paris, 1854–1856. 2 vols.
Du Laurens André Du Laurens. *Historia Anatomica Humani Corporis*. Frankfurt, 1600.
Fabrica Andreas Vesalius. *De Humani Corporis Fabrica*, in *Opera Omnia*, I. Leyden, 1725.
Fallopius Gabriel Fallopius. *Observationes Anatomicae*. Venice, 1561.
G.A. Aristotle. *De Generatione Animalium*, trans. Arthur Platt. Oxford, 1910.
H.A. Aristotle. *Historia Animalium*, trans. D'Arcy Wentworth Thompson. Oxford, 1910.
Kühn *Claudii Galeni Opera Omnia, editionem curavit C. G. Kühn.* Leipzig, 1821–1833. 20 vols. in 22.
Littré Hippocrates. *Oeuvres Complètes*, trans. E. Littré. Paris, 1839–1861. 10 vols.
Massa Nicolò Massa. *Liber Introductorius Anathomiae*. Venice, 1536.
P.A. Aristotle. *De Partibus Animalium*, trans. William Ogle. Oxford, 1911.
Plater Felix Plater. *De Corporis Humani Structura*. Basel, 1583.
Riolan Jean Riolan. *Anthropographia*, in *Anthropographia et Osteologia*. Paris, 1626.

20

THE TEXT

I begin with Jove, O Muses. All things
are filled with Jove.[1]

Lectures on the Whole of Anatomy

by me

William Harvey

Physician of London

Professor of Anatomy

and Surgery

In the year 1616

at the age of 37

Presented April 16, 17, 18
d   J   f

Aristotle, Historia Animalium, bk. 1, ch. 16
There is doubt and ignorance about the internal parts
of man, wherefore it is necessary to study in other
animals those parts which bear a similarity to the
parts of man.

---

[1] This is a line from Virgil's Third Eclogue. Sir Norman Moore has an
excellent discussion of its significance in his presidential address on Harvey
at the Royal Society of Medicine (*Proc. roy. Soc. Med.*, 1915-16, 9, Section
of the History of Medicine, pp. 9-10). Moore traces its origin to the
Phaenomena of Aratus and paraphrases it, "From God all begins: all things
are full of God."

N A T O M Y is that faculty which through in-
spection and dissection reveals the uses and actions of the parts. [There are] five [general] divisions of anatomy: [1] descriptive narrative, [2] use, action and services,[2] whereby [there are], [3] observations of those things which occur rarely and as a morbid condition, [4] resolution of the problems of authors, [5] skill or dexterity in dissection and the condition of the prepared cadaver.

Another [type of] anatomy [is] the ordinary [description] as in this book, of iij venters, [and] the curious and philosophical parts of medicine: 1, physiognomy from the external parts; 2, bones of the skeleton; 3, muscles and ligaments; 4, organs of sensation and of the voice; 5, vessels of the veins, arteries, nerves; 6, similar parts;[3] 7, genitalia, embryo, breasts.

Anatomy [is] philosophical,[4] medical, mechanical.

Since there is division of anatomy, a few things regarding the division. The body is divided: 1, according to the situation, parts: exterior, interior; 2, according to site: superior, inferior, anterior, posterior, here, there, right, left; 3, according to development it has more or less divided into its parts, or it has begun to divide; therefore do not distinguish the anatomy further than nature has divided it.

As the operations and actions of nature are many and distinct, so there are many and distinct parts. The body is the

---

[2] Use may be defined as the aptitude resultant from the form of the particular organ; action, the active motion of that part or organ; and service, the end achieved.

[3] The homogeneous parts of Aristotle, "such as bone, flesh, and the like . . . constituted out of the primary substance" (*P.A.*, 646a:20). Homogeneous parts were of two kinds: simple, amorphous tissue or material, or simple organs composed of a single tissue or material such as a cartilage or bone, or even the heart, since it was composed solely of flesh, (*ibid.*, 646a, n. 5).

[4] Philosophical anatomy was concerned with the causes of structures and the uses and actions proceeding therefrom; the teleological argument.

instrument of the soul, or rather man and part as an instrument having potentiality, like a saw were it able to cut of
its own accord.[5] Hence there are many divisions of the parts
and controversy regarding the part; disputatious authors
differ much among themselves, [cf.] Aristotle and Galen.

WH The part is said to be relative to the whole; therefore since the body [as a whole] is said to be equivocal and
diverse, the part may be in controversy.[6]

And so the body as a whole is called animal in contrast
to plant; living in contrast to [an assembled] mixture, just
as that which is composed and mixed [is] in contrast to
homoieros eteros.[7]

In contrast to plant, since it differs [from plant] in re- 2r
gard to sensation;[8] because it has no part which does not
have sensation. In contrast to a mixture because it has no
part not living; so the contained marrow [is a life-support-
ing substance].[9] Hence more philosophically it is mixed
and composed simply and truly because the soul may be
nutritive, and the nutritive form [is] of a mixture as a triangle in a quadrangle, a fifth in an octave.[10]

WH Thus as the part is a mixture, it is also part of the

[5] *P.A.*, 645a: 14: "As every instrument and every bodily member sub-
serves some partial end, that is to say, some special action, so the whole
body must be destined to minister to some plenary sphere of action. Thus,
the saw is made for sawing, for sawing is a function, and not sawing for
the saw."

[6] Here Harvey appears to be referring to the Aristotelian idea that although the soul is divisible, each part of the body contains the soul as a
whole rather than merely a division of it such as the nutritive soul.

[7] Despite the faulty transliteration from the Greek, Harvey apparently
means "something of a similar nature"; cf. f. 2ʳ.

[8] Aristotle, *De Anima*, 435b:1: "plants because they consist of earth
[only] have no sensation." *P.A.*, 647a:23: "there cannot be an animal
without sensation," but Aristotle does not assert that there is sensation in
every part. *Ibid.*, 647:15: "sensation, then, is confined to the simple or
homogeneous parts," since an organ composed of heterogeneous parts
would be acted upon by several orders of sensibles.

[9] "The marrow also is of the nature of blood. . . . That this is the case
is quite evident in very young animals. For in the embryo the marrow of
the bones has a blood-like appearance. . . . But, as the young animal grows
up and ripens into maturity, the marrow changes its colour, just as do the
external parts and the viscera" (*P.A.*, 651b:20).

[10] Aristotle, *De Anima*, 434a:21: "The nutritive soul then must be possessed by everything that is alive, and every such thing is endowed with
soul from its birth to its death. For what has been born must grow, reach
maturity, and decay—all of which things are impossible without nutrition."

living thing; and as the part is living so the part is sensitive, but not sensitive as a semitone to an octave. Therefore whatever things have been separated by nature differ either through a complete or an obscure distinction, for sensation or power or ease of operation.

WH In whatever way the whole is integrated, the part is a whole. Composed whether: containing, contained, causing movement &c. 2, And so [there is] another division of the parts: [1] philosophical and medical from the end [to be served]; [2] anatomical according to sensation as it relates to the composition.

Sensation may be further [divided into]: similars, HOMOIOMĒROS;[11] dissimilars, heterogeneous. Similar parts for sensation are simple; WH in regard to sensation perhaps no parts of the body can be disregarded. [There is] no part which does not in some way act organically,[12] so that flesh, vital spirit, coction.[13] Use and necessity of the similar parts, because of [1] actions and passions; [2] and as similar in matter or dissimilar. Because of actions: [1] to heat, warm, concoct; [2] to cool, temper; [3] to moisten, soften, lubricate; [4] to dry, absorb, delay, restrain.

Because of the passions [i.e., sensations], as Aristotle 2v [states],[14] through the operations of the senses. So that similar matter but dissimilar [parts], wherefore common to many parts. Special: cornea of the eye, crystalline humor; mucous of the intestines; saliva of the mouth, of the uterus.

[11] I.e., having parts like one another and like the whole of which they are parts; cf. Aristotle, De Caelo, 302a:31 and n. Similar parts: bones, cartilage, ligament, membrane, fiber, nerve, artery, vein, flesh, skin, fat, marrow, nail, hair. Similar parts if subdivided into particles are always of the same substance. Dissimilar parts when subdivided into particles are unlike. The dissimilar parts are the instruments of the soul, since dissimilar parts have definite form. Cf. n. 3.

[12] An organ is a part that can perform a perfect, that is, particular, action. Four orders: 1, consisting only of similars, as muscles; 2, second as those composed of the first, as fingers; 3, made of the second, as a hand; 4, made of the third, as the arm.

[13] The published transcription has the meaningless caro, vitellus, coctio, and although Harvey's writing is difficult to read at this point, it appears to be caro, vital[is spiritus], coctio, and has so been translated.

[14] P.A., 647a:21: "Now as there cannot possibly be an animal without sensation, it follows as a necessary consequence that every animal must have some homogeneous parts; for these alone are capable of sensation."

2, division of similars into liquids, solids. Liquid: blood, sperm, milk, humors of the eye. Viscous: cambium, dew, glue. Vaporous: pituita,[15] bile, mucous, ichor of tears, serum. Of these some [exist] from birth, [some] after birth.

Solids [are] flesh: muscular and fibrous, parenchyma, glandular, [and] special: skin, bladders, gums, pudenda, of throat and intestines, of the tongue &c. The flesh of muscles, gums, parenchyma, glands, is spongy and porous.

Solids [are] soft: as medulla, fat, suet; cerebrum and crystals [of the eyes are] fatty things, sticky and ductile. Medium, as fiber, membrane, covering; [those which are] fissile, thready: vein, artery, skin; fissiles in coverings, nerve, ligament, tendon. Hard [and] frangible as bones and teeth in man; friable as shell in others; flexible as hair, cartilage, nails; and in others [than man] hoof, horn, spine, beak, shell, soft parts, many squames.

3, of all these some [are] of the first order, from which the rest occur or are composed; or are the remaining kinds; 2nd, that which [is] principally from the first and from those remaining.

1st order: sperm, blood, milk; the rest are from these liquids.

2nd order: soft flesh, medium fiber, hard bone. From these others [occur]; the nonregenerated as sperm, menstrual blood. The regenerated.

Dissimilars differ from similars in 4 orders: 1st, from 3r simples, as muscle; 2nd, from the first, as a finger; 3rd, from the 2nd, as the hand; 4th, from the 3rd, as the arm.[16] WH Rather the body.

Philosophical: division of the parts according to the instrument of the soul; according to the divisions of the

---

[15] Galen regarded "pituita" or phlegm as a humor secreted by the brain which left the cranial cavity through the cribriform plate of the ethmoid and passed into the nasal cavity. Vesalius, while not denying the production of phlegm, stated that the pituitary gland received the phlegm and discharged it through the larger foramina near, but not through, the cribriform plate.

[16] This statement from Galen was employed by virtually every writer on anatomy in the second half of the sixteenth century and in the early seventeenth when discussing similar and dissimilar parts. Consequently it would be impossible to determine Harvey's source.

soul; or 2nd, the faculties, from final causes; 3rd, action, use, faculty, function of life, operation. 1, by operations or according to the places where the operations occur in head, chest or abdomen. According to Fernel into regions: private, public; 1st, from the mouth to the hollow of the liver; 2nd, from the hollow to the capillaries; 3rd, of solids.[17] According to Hippocrates: things containing, things contained, those things causing movement.[18]

2, by functions: animal wherefore motion (hands, feet, progression); principal parts: sensitive, vital and natural, wherefore special preparative, concoctive, distributive, perfective functions.

3, by actions and use: organic, instrumental, formative, which produce action; formless having wide use. Whence is apparent what place, what member, what part, particle or part of a part. For there is no part which as an instrument of the soul does not have some action; nor member of which there is no function; wherefore organic parts are not contrary to similars; but (because all in some manner have form) so that they produce for the most part some action or have form; they may be simples and similars in form; according to Fallopius against Fuchs, Fernel, Cardan.

4, by services and manifestations, wherefore necessarily simply, without which no benefit, safeguard, adornment; wherefore principals of first necessity; nonprincipals, ignoble, in which there is less native heat, less spirit and worth.

Other parts in addition; sympathetic; through similarity 3v of kind, association of service, communion of vessels.

These things generally, of which there are many observations, many doubts and problems, but the particular

---

[17] *Methodus Medendi*, in *Universa Medicina* (Paris, 1567), p. 359. Fernel's concern was with a division based upon what he considered to be the routes of expurgation. The public regions were 1, from gullet to middle of liver, which included stomach, mesenteric veins, spleen, and pancreas; 2, middle of liver into the hairlike veins of particular parts, including the convexity of the liver, vena cava, aorta and their parts from the axilla to the groin; 3, muscles, membranes, bones, and mass of the body. The private regions with "their own superfluities and special passages for expurgations" were such as the brain, lungs, kidneys, and uterus.
[18] Hippocrates, *Epidemics* bk. 6, sec. 8, par. 7; Littré, V, 347.

work of anatomy. [According to] Augustine all parts of nature with adornment may be necessary, but not all necessary which are for the sake of adornment; wherefore nature more solicitous of adornment than of necessity, as this apostle understands from the resurrection.[19]

Canons of general anatomy

1. *Shew as much* in one observation *as can be*, such as the whole belly; or what things occur in some entire portion [of the body]; then subdivide according to sites and connections.

2. To demonstrate the things appropriate to that cadaver. [Mention] new things or [those] newly discovered.

3. *To supplye only by speech that which cannot be shown on your own credit and by authority.*

4. *Cutt vp as much as may be* in the presence [of the spectators] so that skill may illustrate the narrative.

5. To recount one's own and others' observations, to confirm one's own opinion or to give consideration [to the anatomy] of other animals; according to the Socratic rule: *where it is farer written.* Therefore [refer to] foreign observations.

1, on the causes of diseases: especially useful for physicians.

2, on the variety of nature [and other] philosophical [ideas].

3, for refuting errors and therefore solving [problems].

4, on uses and actions for discovering merits and therefore [have] collected passages. For the goal of anatomy is understanding the necessity and use of the part. Especially [is this true] for philosophers who thence know those actions which are required for

---

[19] Augustine, *Of the Citie of God . . . Englished first by J. H.* (London, 1620), bk. 22, ch. 24, p. 848: "There are some parts of the body that concerne *decorum* onely, and are of no use: such are the pappes on the brests of men, and the beard, which is no strengthning, but an ornament to the face, as the naked chinnes of women (which being weaker, were otherwise to have this strengthning also) doe plainly declare."

each [part], and display to physicians likewise who thence [know] the natural constitution [and] the rule of classifying the patients and therefore what ought to be done for the diseases.

6. *Not to prayse or disprayse* [other anatomists]; *all did well and behowlding* to those who were in error [solely] because of chance.

7. *Not to dispute* [or] *confute* otherwise than by demonstrable arguments because more than iij days are required [if this be done].

8. *Brefly and playnly* [state your case] *yett not letting pas* 4v *any one thing vnspoken of which is subject to the vew.*

9.[20] 10. *Not to speake anything which with outt the carcase may be delivered or read att home.*

11. For not too much concern may be attached to single dissections and time does not permit [repetition].

12. *To serve in their iij courses according to the glass.*[21]

1st lower venter, *nasty yett recompensed by admirable variety.*

2nd *the parlor.*[22]

3rd *divine Banquet of the brayne.*

In the account of anatomy 5r

In all parts consider: temperament, what results, what happens.

WH Organic parts 5: 1 site, 2 shape, 3 quantity, 4 motion, 5 division.

Site or circumscription: before, behind, above, below, here, there, originates, terminates, occurs, passes away. It is joined by connections: fibrous, membranous, nervous, ligamentous. Vessels: veins, artery. Whence agreement of parts, relation of vessels.

Shape: of the whole [and] with skillful similitude, proportion of the parts, conformation, beauty; of the surface:

---

[20] This numeral is missing from the published transcription.
[21] I.e., keep within the limits of time.
[22] This means the thorax. The parlor was the inner chamber, a term also used by Harvey; cf. f. 8ʳ.

PAGE 5ʳ OF THE NOTEBOOK

smooth, rough, hairless, hirsute, even, uneven; of the passages: entering, emptying.

Quantity: separate parts: number; continuous parts: length, width, depth, size, capacity.

Motion: local; quantity, augmentation, diminution, generation; kind according to diseases, habit, age.

Division into parts: according to site, before, to the rear; according to position, outside, within; according to composition, wherefore it is apparent.

Of substance: sanguineous, fleshy, sinewy, membranous, of skin; from these [arise] temperament, strength, force, sensitivity, color, generation.

Individually: what problem results; what denotes as a sign; and because of which. Necessity, use, action, worthiness.

Individual observations: considering the approach both special and foreign from books. In the same species: age, sex; diseases: occur, expelled, carried away; and habit. Different [species] but have [similar] parts of it: in fowl: riverbank, terrestrial, aquatic; in fish, serpents; in oviparous quadrupeds; in viviparous quadrupeds.

In the account of similar parts 5v

Substance: color opposed to or derived from blood; thickness, thinness, hardness, softness, density, rarity. Whence temperament, strength or fragility, sensitivity.

From temperament [is determined] shape or site of the parts.

Motion: active, passive. Amounts: augmentation, diminution; generation: material, proficient.

Kinds: color, in thickness, in complete hardness, in density &c; temperament, strength.

Passions: according to nature, contrary to nature.

In the actions of the parts 6r

Since the purpose of anatomy is to know or to recognize the parts and to know them through their causes, and these in all animals, for the sake of which and therefore for this reason, 1, action, 2, use.

1, action, the active motion of which is an effect, is called function; but in matter, operation. Operation and function by themselves and with the other parts; [the organic part is] principal or assisting; [principal organic parts are liver, heart, brain; remaining parts] assisting in accomplishing or in preserving [the action].

2, uses and ends: mediate. According to likeness: hot, to make hot; cook, to foment; cold, to make cold, to temper; humid, to lubricate, to smooth, to soften, to blunt; dry, to make firm, to strengthen. For color of blood, active temperament; hardness, softness, passive temperament; density, lightness, heaviness; thickness, strength, fragility. For organic parts: shape, quantity, site, composition. The ends follow use and actions.

Final [ends]; for being wherefore necessity; for well-being wherefore worthiness; for safeguarding, sine qua non, wherefore necessity; for adornment.

Excellence according to ends, for [the result of] many things. Necessity sine qua non for being; worthiness, well-being and perfection, safeguarding, adornment. And this is elicited from my own observations and from opinions of others, from accounts not in complete agreement, considering that part either in general or in proportion.

Little known to us; those things necessary; permitted contrary to the causes; those in which sperm, organs, matter.

[Blank] 6v

The body [divided] into trunk, limbs; wherefore some 7r [parts] symmetrical, [others] asymmetrical.

Limbs: for the sake of motion wherefore movements of the parts [occur] from junctures of bones and muscles. Composed of legs and feet [or] of feet alone. 4 actions and uses: progression, local motion, standing, sitting. Those which lack feet [possess] 3; wings, length of body in which the longer it is the smaller the feet, e.g., *lizard;* winged fowl in which the larger the wing the smaller the feet.

31

Grasp of hand and especially for conveyance of nourishment; wherefore children desire to carry all things to the
mouth. 3 uses: caring for the body; defending one's self;
the skills of the parts existing before the parts so that the
shape [is adapted] to the nature of the intellect.[23]
Those which lack hands compensate for their use by
length of neck and beak; by the tongue [in] bees, dogs; by
the trunk in the elephant; butterfly; by the feet in apes,
parrot. WH I have learned to grasp with my feet.
Al bur.[24] Those who in the judgment of Vitruvius are
considered to preserve natural proportion. In height: of
tall stature, a sign of liberality; of compressed [stature],
short; compact in width and depth; expanded or deficient
in some. [All these are] brawny, muscular, squarely built,
fat, thin, skinny or *meagers*.
Those slightly exceeding, those slightly deficient [in
natural proportion]; exceeding: giants, heroes, as royal
drones; deficient and of trifling size, troglodytes of whom
I know nothing in regard to the body.
All parts of the hand reach to the buttocks; the legs sustain and carry the weight [of the body]. Unequal because
they do not preserve a constant size: [1] long limbs, slender body, long shanks, *gangrels*,[25] *long Harry*;[26] a kind of

[23] Cf. Galen, *De Usu Partium*, I, iii, Daremberg, I, 114–115; in which
it is asserted that the form of the hand is the result of man's need as directed
by his intellect. "It is not by his hands but by his reason that man has
learned skills."
[24] In view of Harvey's adjacent reference to Vitruvius's canon of proportion, and despite his spelling, the reference here must be to Albrecht
Dürer's studies of proportion, *Hierinnen sind begriffen vier Bücher von
menschlicher Proportion* (Nuremberg, 1528). Presumably Harvey would
have employed one of the several later Latin editions.
[25] The published transcription erroneously gives "gengreti." The word
gangrels, now obsolete, is associated with "gangling." It is formed by
analogy with "mongrel," and means a lanky, loose-jointed person. Cf.
Hadrian Junius, *The Nomenclator* (London, 1585), pp. 449–451: "A long
gangrell, a slim; a long tall fellow that hath no making to his height."
[26] *long Harry* (not *Herry* as erroneously in the published transcription)
was the nickname of Henry Saville, fourth son of Thomas Saville, of
Banke, Yorks. He was born in 1568, went to Merton College, Oxford
(M.A., 1595), and was licensed to practise medicine on November 18,
1601. He died on April 29, 1617, and was buried in the church of St.
Martins-in-the-Fields. Commonly known as "long Harry" because of his
height, he was an eminent scholar, especially in painting, heraldry, and
antiquities. *See* J. Foster, *Alumni Oxonienses* (1891), IV, 1319.

spiders, fowl, ostrich, stork, crane; [2] very large body, short limbs, dwarfs. Infants in first months: hands scarcely extending to chest; feet at umbilicus; appendices as in frogs; tortoise; all quadruped animals, also the horse.

They excel in those operations according to the degree of their natural fitness, as the industrious and ambitious in their hands, and just as Alexander was able to stand upright on his feet for a long time. Also birds, as the swan, sleep on one foot. [According to] Aristotle greater fluidity of the bowel in one body;[27] but, on the other hand, in another body, providing [there are] many concoctions, such are distinguished by fat, great spirit and tenacity; they are unable to stand for long on one [foot]; *waddle like a puffin;* wherefore "Solomon" Ecclesiasticus 19:28, gait, laughter and attire.[28] For some employing measured, regular step walk [in the manner in which] the executioner approaches; others move about here and there in a disorderly confusion like ducks; wherefore those of disorganized habits, pertinacious *dogged fellowes.* Wherefore 3 kinds of dwarfs: [1] very small, proportioned pigmies; [2] dwarfs, sm[all] body, shapeless, *vgly;* [3] those in whom the spine is curved, humped, limbs rather long; *gibber Gobbo Nang.*[29]

The trunk is divided into regions: anterior, posterior;

[27] This seems to be a reference to *P.A.*, 674b:29: "Other birds there are, such, namely, as have long legs and live in marshes, that have none of these provisions, but merely an elongated oesophagus. The explanation of this is to be found in the moist character of their food. For all these birds feed on substances easy of reduction, and their food being moist and not requiring much concoction, their digestive cavities are of a corresponding character."

[28] Harvey's citation is incorrect, since this phrase is Ecclesiasticus 19:27, which reads in the Douai version of the Vulgate: "The attire of the body, and the laughter of the teeth, and the gait of the man, show what he is."

[29] *Gobbo,* Italian word for humpbacked. Certain edicts of Venice were proclaimed from a figure of a humpbacked dwarf known as Gobbo. We are reminded of Launcelot Gobbo in the *Merchant of Venice.* Nowhere else is the character alluded to in this place as being humpbacked, nor is there any mention in travel books of Shakespeare's time of the Gobbo of the market place. *Nang* is possibly a misspelling of the French word *nain,* meaning humpbacked. Cf. S. Weir Mitchell, *Some recently discovered letters of William Harvey* (Philadelphia, 1912), pp. 9–11.

wherefore the special sense of vision before the eyes.[30] All things are united in the heart wherefore it is the chief place;[31] the heart the principium of man as the center of a circle; above [the heart] nourishment, wherefore [a contrast] to plants; below, excrement; above, downward; for man upward, for animal proportionate to the amount of [bodily] heat. Right, left; whence and whither motion. Horses, hares prefer to lie on the left side; defect of man from childhood; *foot-ball;*[32] they sleep on the right side; *turne on the too: lepp: hop* [on the right foot].

Superior, hotter and more worthy parts [on the] right; left &c., [colder] and [more] blood [required] for them; wherefore ulcers are healed more swiftly on the right;[33] the contrary for ulcers, tumors, cacoethes[34] [on the left side], but those things which are symptoms so that more frequently ulcers, bubones on the left side, downward, internal, privately.

[The trunk is divided into] parts: [1] KEPHALE, head and neck [are above the trunk] for so it appears: *Agypt,*[35] Aristotle,[36] Ruffinus book 3;[37] [2] from the first vertebra to the vertex for which a large chapter; [3] thorax to the clavicle to the pubes and exit of excrement; [according to]

[30] *P.A.*, 656b:30: "The organs of vision are placed in front, because sight is exercised only in a straight line, and moving as we do in a forward direction it is necessary that we should see before us, in the direction of our motion."

[31] Later, in the dedication of his *De Motu Cordis* (1628), Harvey was to say: "The animal's heart is the basis of its life, its chief member, the sum of its microcosm; on the heart all its activity depends, from the heart all its liveliness and strength arises."

[32] Meaning either that the ball is normally kicked with the right foot or that development and use of the right leg and side is the abnormal result.

[33] There is no scientific basis for this statement, which is not true.

[34] Galen used the term to signify an incurable ulcer.

[35] Crooke, p. 63: "The Egyptians [divide the body] into the head, the necke, the chest, the hands, and the legs."

[36] *H.A.*, 491a:26: "The chief parts into which the body as a whole is subdivided, are the head, the neck, the trunk (extending from the neck to the privy parts), which is called the thorax."

[37] This appears to be a reference to Rufus of Ephesus, *De Appellationibus Partium Corporis Humani*, III, i-v, in *Medicae Artis Principes . . .* (Geneva, 1567), coll. 119–122, where Rufus describes the divisions of the body according to osteology.

Aristotle more visible in birds[28] &c.: STÉTHOS, chest, wherefore in Hippocrates it is called the upper venter, aphorism 38 section 7;[39] GASTER, abdomen.[40]

The trunk is commonly [divided] into venters, not without reason. Upper [venter]: head, brain; domicile, shrine, citadel of the spirit where the animal senses, where the intellect and reason [reside]. Middle [venter]: chest, heart; inner chamber, dwelling where [lodges] the source of heat, vital [spirit], anger, emotions, respiration. Lower [venter]: abdomen, liver; kitchen [or] shop where coction of nourishment, natural nutrition, generation [take place].[41]

Venters: 1, distinction: distinct [in] bees, wasps; as in animals; connected in fish; in crustacea in middle manner: head, and rest partly diaphragm, partly not; 2, order; 3, dignity; 4, difference in constitution; 5, signs of physiognomy.

2, Order: [1] senses in exact location in the anterior 8v and superior part of the brain. Use: to distinguish between that which is nourishing as opposed to nonnourishing; and to give rise to motion and flight from harm. [2] in the lower part, as in a building, [is] the kitchen [for] cooking of food. As brain and heart [are above], so foul and smoky vapors of cooking [are below] lest things be harmed. [3] in the middle part which primarily is the foundation. In the middle the virtue and foundation of each; here nutrition and generation, there [above] senses and intellect.

3, Worthiness, excellence. Simply for the existence of what is here; sine qua non so that that which is here may be perpetuated [1] for individuals, [2] for species. Sensations for the well-being of what is here. And so it is proper that the head is more worthy and honorable than the chest. But

[28] *P.A.* 693a:24: "The upper and under sides of the body, that is of what in quadrupeds is called the trunk, present in birds one unbroken surface."

[39] The published transcription incorrectly gives aphorism 30—a surprising error, since reference to the text would have indicated the reading as meaningless. Harvey is indicating that Hippocrates called the chest the "upper venter," so dividing the trunk into two venters.

[40] So given in Hippocrates, *Epidemics*, bk. 6, sec. 6, par. 4.

[41] A rather apt simile which was used by a number of authors.

35

it is more necessary that the chest be first and fundamental. In second place the abdomen because like earth; internally in animals as outwardly in plants.

4, Difference in constitution. Brain for worthiness; bony [covering] for security, safeguarding. The liver in distention by nourishment, by flatus, by uterus. Fleshiness of region of the heart for protection; fleshy for distension of the lungs; osseous for safeguarding.

5, Hence many signs of physiognomy which the true art may neglect. [1] head: large, small. [2] broad chest and dearticulated and perhaps according to the spirit; for males thus [show] signs of it. Amplitude of nostrils in horses and width of chest [related] to breadth of shoulders. [3] venter: insensitivity of the voracious; spirit in the belly venter; amplitude of mouth, extends for a half yard; those things which have been provided for gullet and stomach which exist one and a half feet distant from the mouth in animals; and in fish; compressed place for the senses of life; infinitely voracious; [42] equilibrium: I determine the male by the chest, the female by the gut. Hence longer from [ensiform cartilage] downward [to] the umbilicus than above from the root of the neck to [ensiform] cartilage; shorter life because the heart is burdened by excrement.

---

[42] *P.A.*, 675a:20: "The whole tribe of fishes is of gluttonous appetite, owing to the arrangement for the reduction of their food being very imperfect."

→

[43] The word clavicle is derived from *clavis*, but the bone is quite unlike any key. However, it is rather like the Roman hoop-stick (*clavis trochi*). It seems more reasonable that it got its name from being the key-bone joining the scapula to the sternum.

[44] I.e., heartburn.

[45] A cardiogmos is a burning pain in the epigastrium and has the same meaning as cardialgia.

[46] Galbanum is the gum from an evergreen plant grown in Syria and Africa (*bubon galbanum*). A plaster made from it was formerly applied to the abdomen in hysterical conditions or to allay pains in the uterus after delivery.

[47] Harvey is in error, since Colombo (p. 100) says, "The vertebrae of the loins are five in number."

[48] A swelling in the inguinal region, usually associated with inflammation of the inguinal lymphatic glands, especially in plague.

[49] "Pecten pubis" means the symphysis pubis. The word pecten was used by the medieval anatomists to mean either the pubis, the metacarpals, or the metatarsals. Originally Pliny used the term for the pubic hair; hence its application to the pubis (Charles J. Singer and C. B. Rabin, *Prelude to Modern Science. . . .* Cambridge, 1946).

NŌTON, the back from neck to

he thorax is vided; useful r physicians motions are sible from e site of the fected place	Posterior part	Metaphrenon	Emplaster of the stomach Introflection of pregnant women I have seen and cured dislocation from swellings with resolution of the lower part
		RACHES, dorsal spine	
		Scapulae, ŌMOPLATAI	
		Loins, in English *Reyns*, kidneys higher	
		Coxendix, region of ischiae, wherefore sciatic pain	
		Region of sacral bone, medically attachment of straight intestine (rectum)	
		Region of coccyx	
		Nates in form of cushioned seat; in men fleshy, in apes calloused	

oportion of est to venter ourth as 3/4, ur thirds; of e chest to e head 2/3, a th; of the est to the wel 2/1 an tave	Anterior part	Chest from clavicle to ensiform cartilage		Sternum wider in man, and in those in which wider more suited for vomiting. In other animals keeled, *like the keel of a ship*, wherefore birds [use] as a foot when they sleep
			Jugulum	Above, keys (claves),[43] clavicle Below, axilla
				Pleurai and notai which do not reach to the sternum
				KARDIA, the little trench of the heart where cardialgia[44]
				Hence the heart in some cardiogmos[45] and the ignorant demonstrate the heart to be there; thus *att hart*
				Breast, nipples

	Inferior venter is	3 regions of the venter	Epigastric	Stomach cavity; *hole of ye stomach* Hypochondrial subcartilage right left
			Gastric	OMPHALOS, root of venter wrinkled skin of aged emplaster of galbanum[46] [over] uterus Ilia, loins, *flanke* Colombo [says] 6 vertebrae in length,[47] some 4, commonly 5
		Its circumscription such	Hypogastric uterus and bladder	Bubones, wherefore bubo[48] in groin Pecten pubis,[49] wherefore more prominent in pubescence in women for easy parturition
		Internal parts of which later	Containing Contained partly containing, partly contained	
		Exterior	Common	Skin, epidermis, fat, suet, membranous, fleshy
			Special	Muscles of abdomen Umbilicus

Skin. 10, The whole rampart and bulwark [of the body].   10r
A complementary body holding all things together. Some
[uses]: emunctory, medium of touch, organ of touch.
Some say X [that the skin] carries down nourishment from
the brain as X proved in the tree. Substance of the fleshy
parts whence nervous flesh[50] &c. Those things which result.

3, The most temperate part [of] the skin [is that] of the
fingers of the hand, wherefore the sense of touch as according to Galen[51]—for example: hot and cold water [felt al-
ternately seem] of medium quality to the touch of all. Man
differs from man in other parts, according to temperament,
softness and hardness of skin.

[Skin] varies in thickness, thinness, hardness and softness. According to Aristotle it is thinnest in man in relation
to size.[52] But these differences seem according to age, sex,
temperament, complexion, nature of diet, of air. Wherefore in some thick, [in others] thin, rare, dense, soft, hard.

Corium: of very fatty humor; it seems more aqueous [to
the touch] than goldbeater's skin.[53] WH *Tougher* where
thinner Δ on the external hand; [skin of hippopotamus so
thick] the fore part *turne a lance*.[54] In all the nature [of the
skin is] earthy because most frequently exposed to injuries; the humor evaporating the solid earthiness [re-
mains]; not moist, sticky mucous as appears in glue.[55] In
tabes[56] very thin, not the loose rough [skin] of emaciated

[50] Crooke, p. 87: "Galen is of opinion that the Skin is absolutely temperate, because it is of a middle nature betweene bloody and vnbloody, whence
it is called Neruous Flesh and a Fleshy Nerue." *Ibid.*, p. 88: The skin
"may therefore be called a Fleshy Nerue, or a Neruous Flesh, because it
hath a middle nature betweene Flesh and a Sinew: For it is not vtterly without Blood as a Nerue, nor so abounding with Blood as Flesh."
[51] *De Usu Partium*, II, vi; Kühn, III, 110; Daremberg, I, 181.
[52] *G.A.*, 517b:27: "Of all animals man has the most delicate skin:
that is, if we take into consideration his relative size." Cf. *ibid.*, 785b:5.
[53] Harvey uses the term *cutis* at the top of this page for skin. He also uses
corium, here translated as *dermis*. In a modern sense the corium is the layer
of skin beneath the epidermis. It would seem that Harvey was using it in
this sense, for he compares it with goldbeater's skin.
[54] Du Laurens, p. 215.
[55] *H.A.*, 517b:28: "In the skin or hide of all animals there is a mucous
liquid, scanty in some animals and plentiful in others, as, for instance, in
the hide of the ox; for men manufacture glue out of it."
[56] The term is used to mean any wasting disease or cachexia.

persons; as of the elderly; [according to] Galen at first loose;[57] varies in different parts [according to location and use].[58]

Thickness [of skin] of the head &c.[59] In some diseases dry and taut. *Hidebowned.*[60]

Signs of physiognomy; from softness and hardness [indication of humid and dry temperaments respectively]. Innate inclinations; those whose skin is as thick as horse's *will eate meat;* healthy, slender, agile, sprightly; tougher and thick, boldness, sound health of body, but of slower and violent wit. Thin [skin]: benign nature, apparent gentleness; exquisite sensitivity wherefore internal sensation. The tougher [the skin] the duller is the sensation.

4, Color as in substance between flesh and sinew. Cause 10v of color in those in whom the skin [is thin]; the nose; hairs in the snout &c. In man according [to the condition of] the skin, so the hair [is thick or thin]; greasy [excrement from skin] as in morphew;[61] [some] skin causes hair to fall out; [hair] grows in unlikely places; *the boy abowt Holborn bridg Beard* on the one cheek.[62]

[57] *Comm. III in Hippocratis Epidemiorum lib. VI;* Kühn, Vol. XVIII, pt. 2, p. 90: "Some wrinkling appears in the bodies of the young."
[58] Crooke, p. 73: "According to the diuers vse of the parts, it is either softer or thinner, as in the Face, the Yard, and the Scrotum or Cod; or harder, as in the Necke, the Backe, the Legs & the Soales of the Feet; some of it is in a middle temper between hard and soft, as in the Palme of the Hand, and especially in the Fingers ends, because they are ordained to apprehend with. . . . "
[59] Harvey might have been thinking here of one or both of the following beliefs of his age. Since the head lacked fat, the brain necessarily required the protection of a thicker skin, and according to Crooke, p. 73, "So some part of the Skin is exceeding thicke, as in the Head"; furthermore, the greater hairiness of the head was the result of the thick skin of the head, because, as Crooke remarks, p. 69: "the haires of the head are the longest of all the body, because . . . the skin of the heade is exceeding thicke, yet rare and containing much moysture."
[60] Cf. Gervase Markham, *Markhams Maister-peece* (London, 1623), p. 83: "This disease which we call hide-bound, is when a horses skinne cleaveth so hard to his ribbes and backe, that you cannot with your hand pull up or loosen the one from the other. . . . The signes, besides the cleaving of his skin, is, leannesse of body, gauntness of belly, and the standing up of the ridge-bone of his backe."
[61] The scurf, dandruff.
[62] There are numerous cases on record of anomalous growths of hair, many of them attributed to a hairy naevus, of which this might have been an example.

39

Man has hair on the skin where it is lacking to animals and on the contrary, lacks it where it exists in others. Some in the groin, beard on the chest, [hair on] the anterior [body], *mustach*. In other animals more posterior except in birds.

5, Generation spermatic because [some hair grows only after puberty]; X WH Δ in the first conformation [of the child] none on eyebrows, cheeks or lips. It is generated as in a drying pottage of meat[63] &c.; in serpents, crustacea [skin too dry to grow hair];[64] in trees, man [too much moisture prevents growth].

In man the more it is rubbed away [the more it grows].[65] The contrary in dogs and apes, but I believe in those [it will also grow]. *spurrd Hackneys, saynts* on the knees;[66] Plautus: brawn of boar;[67] when the skin has been lost thick cicatrix [results]; wherefore to cauterize with needle to produce cicatrix; useful for luxations, ruptures &c.

Vessels, veins 6: 2 from the jugular, 2 from the axilla, 2 from the groin;[68] according to Oribasius the kings of Persia [employed] skin for windows;[69] according to Aeneas Sylvius *Ziskas scin made* [into] *drum*.[70] Mr Havers

[63] I.e., moderately dry to hold the hair at its root and to maintain the hole pierced by the hair.
[64] Skins of snakes and crustacea too dry to grow hair (Galen, *De Temperamentis*).
[65] I.e., as callous.
[66] The published transcription has, in error, "*spurred Hackney saynts genibus.*" Spurred Hackneys have bare patches on their flanks where the hair will no longer grow; *saynts* is obviously an example of Harvey's humor and perhaps a hit at the Puritan "saints" who from much kneeling had hard and calloused skin on their knees.
[67] *Persians*, II, v, 305.
[68] Crooke, p. 88: "The common opinion is, that it ariseth from the dilated ends of the Veins, Arteries and Sinewes, because it euery where feeleth, liueth, and is nourished: now life is communicated by the Arteries, nourishment by the Veins, & Sence by the nerues or sinewes. For my owne part I doe not deny but that many vessels are carried unto and doe determine in the skin from the Axillary, Iugular, and Crurall veins, many small Surcles, and as many Arteries bearing them company . . . but I am not resolued that skinne is wouen together of their threds."
[69] Most likely taken not directly from Oribasius but from Du Laurens, p. 214: "wherefore we read that a certain king of the Persians once used it for making windows."
[70] Aeneas Sylvius, *Historia Bohemica*, ch. 46: "It is said that when he was ill he was asked where he wished to be buried upon his death. He ordered that his skin be stripped off, the flesh cast aside for vultures and

40

pricks of the poths appeare in the scin of deer in Croxton.[71]
6, Shape; in a mirror infinite lines and wrinkles &c. 11r
3, In withering age &c. In some more [in some less];
greyne; where rouge clings; wherefore horses are more
correctly treated [when curried]; skin in folds. Wherefore among the Venetians *god make me fatt* [and un-
wrinkled]; wherefore more wrinkled among the elderly
&c; wherefore *the scin not for paynting for* &c.; likewise
sordes [collects] in the wrinkles; wherefore if they know
[they cover with] a pinch of semola; vna suppa divina.

It is clearly perforated in 8 places &c., but perhaps not
at the anus; but it is perforated at the nails, eyes, umbilicus; pores &c.; so that hypericum [expels];[72] for sooty
dross &c.; sweat an indication. Hair; in the pores wherefore the pores oblique as the site of the hairs, hence regarding friction NB. WH Wherefore I have scraped the porosities against the nails. [According to] Hippocrates easy in
those of more rare skin &c.;[73] wherefore according to
physicians, as [according to] Aristotle obstruction to all
[external harm] and cold air.[74]

In those in whom too open [according to] Aristotle a
kind of bloody excrement,[75] and [according to] Galen,

---

wild beasts, and a drum made from his skin; with it in the forefront when
battles were waged the enemy would take to flight as soon as they heard
the sound of the drum."
[71] Croxton Park, which is about two miles north of Thetford in west
Norfolk, was in Harvey's time one of the estates of Thomas, Earl of
Arundel, and was well stocked with deer (*see* F. Blomefield, *Topographical
history of Norfolk*, V, ii [London, 1805], 150–151). Mr. Havers was probably a bailiff or steward, as a man of that name is recorded as being a
trusted servant of the Duke of Norfolk (Thomas's grandfather) in 1572
(*see* Hist. MSS. Comm., *Salisbury Papers*, ii [1888], p. 5). "*Poths*" is an
obscure word which has not been traced. It may be a variant spelling of
"potes," a dialect word referring to sticks or poles or anything similar used
for poking or pushing.
[72] The name given to St.-John's-Wort, the leaves of which produce an
essential oil like turpentine.
[73] *Aphorisms*, V. 69: "In women rigors commence especially at the loins
and travel from the back to the head; in men also rather in the posterior
than in the anterior parts of the body as from the forearms and thighs;
there men have rare skin as the hair which grows on them demonstrates."
[74] *P.A.*, 658a:19.
[75] *P.A.*, 668b:6: "Instances, indeed, are not unknown of persons who in
consequence of a cachectic state have secreted sweat that resembled blood,
their body having become loose and flabby. . . . "

41

Com[mentary][76] de Fracturis. [Du] Laurens testifies to have observed exudation of blood in the English sweat;[77] *doctor Gilbert* &c.;[78] blood by thinness and heat of the open pores wherefore the same in some who have taken cantharides.

Shape of the skin is that of the body, wherefore *the best fashion to lep to run to doe anything stripp to ye scin. Fashion is but* a redundant covering, a fantastic arrangement.

7, Motion: I have observed contrary porosities. Espe- 11v cially in regard to the emotions; in fright *see the divil* &c.; the hair stood up in horror; chill from a story, *cowld, a goose scin.* It is apparent: in the hair, [in] *a goose scin.*

Causes: by fear, horror, fever, chill, wakefulness, disease. In other animals and birds from disease, from horror; when had stood in water; in dogs, in those fearing other things. In certain ones by voluntary motion: *porcupin hedghog turkey cocktoo, ruff Bird in ye Ballat.*[79] In men: in wakefulness; early in the morning, *lord how you look as gamesters,*[80] *sick leane dog, Begger sick;* the hair arose in horror. Hence use of mordicative medicines against dry ulcers.

8, Sensitivity: in man especially for pudendum, root of penis, nipples where &c.

Two notable nerves, *Nan gunter*[81] &c.; I believe had pro-

[76] Here the published transcription gives the meaningless phrase "Corii de fracturis."

[77] *Du Laurens,* p. 214: "Sometimes in them the imperceptible passages open so that pure blood flows out, as in the English sweat, and Galen in his Commentary on the Book of Fractures observes."

[78] William Gilbert (1540-1603) was physician to Queen Elizabeth I and author of the celebrated work *De Magnete* (1600), but he does not seem to have left any observations or comments on the English Sweat, the main outbreaks of which occurred before his time. It is possible that Harvey meant to write "doctor Caius," whose monograph on the English Sweat (1552) is a classic of clinical description. Both men were presidents of the College of Physicians.

[79] The ruff-bird (*Machetes pagnax* L.), is described by Sir Thomas Browne, *Works,* ed. Wilkins (London, 1910), III, 319. During the spring months the male has a handsome ruff of feathers round its neck which it can erect or depress at pleasure. "Ballat" is an old spelling of "ballad."

[80] This phrase sounds like a quotation, but its source has not been found. It refers of course to the exhausted appearance of men who have spent the night gambling.

[81] As Hunter and Macalpine have shown ("A note on William Harvey's

42

duced a callous; *the mad woman pins in her arme*; *Mary pin her cross-cloth: beginning with the* incantation *as* boys in palm of the hand.

Hence a certain anatomist accuses Aristotle, and a certain one excuses because he said that the skin does not have sensation, especially in the head where indeed man requires [it].[82] **WH** Aristotle regarding the skin of all animals as wholly [devoid of sensation]; for usually they are not sensitive, such as fowl, quadrupeds, the oviparous, serpents &c.

9, Connection. In many places inseparably [connected] 12r with the panniculus carnosus as in sheepskins;[83] especially in fowl and plants; looser at the *elboe* and ankle because of [need for] motion; [more closely connected] at the ribs except in the fat when [it is more easily removed]; separated from the flesh it is reduced, stretched, loses its color as in the incurably ill &c., which not the original [tone].

10, Site. In the outermost part of the body where, [as] commonly [believed it arises from the dilated ends of veins, arteries and sinews].[84] Rather, I believe as in fowl in which heat &c. Hence some arises from the extremity of the vessels. The universal emunctory; witness the effluvia and sweat. Weakness: easily [burned] by sun and wind [in] *one jorney*. 1, likewise as in many ulcers infinite diseases; 2, it is easily colored from the offending humors as jaundiced, pallid &c.; generalized ulceration &c.; stig-

---

'Nan Gunter,' " *J. Hist. Med.* [1957], 12:512–515), Anne Gunter, who falsely claimed to have been bewitched, was a celebrated case in London in 1604–1605 and the subject of official investigation by the Royal College of Physicians, the Star Chamber, and James I himself. The same authors also discuss the allusion to two similar cases. See also their paper "William Harvey: his neurological and psychiatric observations," *J. Hist. Med.* (1957), 12:126–139.

[82] *H.A.*, 517b:32: "The skin, when cut, is in itself devoid of sensation; and this is especially the case with the skin on the head, owing to there being no flesh between it and the skull."

[83] Unlike the epidermis, since this "in tanning will fall away, whence comes the holes in Sheep-Skins when they are made into Parchment" (Crooke, p. 88). In man the skin is firmly bound down to the underlying tissues in certain sites, notably the palm of the hand and the sole of the foot. Cf. n. 542.

[84] "because it every-where feeleth, lieth and is nourished" (*Crooke*, p. 88).

mata, dermatitis. Cause of weakness because in a way the expansion occurs outwardly far from the heart.[85]

It is divided: skin; epidermis, outward skin, dried surface.

Epidermis. It is generated continuously &c.; is regenerated like cappo di latta [the scum of milk]; whence epidermis, *scurfscin;* wherefore [the skin] of the [naked] infant [of reddish color];[86] and we see it very dried out. 2, analogy on all sides, of trees, crustacea, serpents, of old age. Man with dandruff, *meale* for birds, animals rolling about and scratching; in a horse *curried, the bird att St. James bemealing as a man* [with] *scarlet* [fever].[87]

3, Connection: not separate; whence X according to 12v anatomists as if a part in itself; for that which is separate &c.; yet separable in tumors, phoenigmi ignei.[88] *Burning scald,* wherefore to remove with cosmetic; not in dead person; Problemata of Cassius; X he attempted.[89]

[85] "The weaknesse of the Skin proceedeth . . . by reason of the scituation and the Vessels. For the greater Vessels because they are neerer to the Fountaine are the stronger, and the expelling vertue of the inner-parts more powerfull: Whence it is, that the inward parts expell their superfluities into the outward, and the greater vessels into the smaller vessels of the habit; so that the Skin becommeth weaker, because the expelling faculty is withinward & stronger, and layeth all the burden vpon the Skin" (*ibid.*, p. 87).

[86] "For whilest Nature in the generation of the Skin mingleth Blood with the Seed, a moist vapour of the Blood foaming or frothing vp, and driven forth by the strength of the heat, is condensed or thickned by the coldnesse of the Aire, and turned into a Cuticle or Scarfe-skin, for so I thinke we may properly call it. And this is the reason why in Infants new-borne, the whole Skin looketh red, the Cuticle not being yet formed for want of cold Aire; or at least not sufficiently condensed, as it is after a short time; the Aire thickning the creame or froth of the Blood, as we see in Gruell or boyling of Past or Starch, a Skin, filme or phlegme gathered together of the vaporous froth that ariseth from the thicke moisture which is by the cold Aire condensed" (*ibid.*, pp. 71–72).

[87] St. James's was the site of the royal aviary, where Harvey had apparently seen a bird shedding "scales" as does a man recovering from scarlet fever.

[88] Phoenigmi were substances that produced a local reddening when applied to the skin. Strictly, they were called *epiplastica phoenigmi.* More caustic substances were called *epiplastica vesicatorii. Phoenigmi ignei* may in this sense mean the reddening due to fire.

[89] *Cassii medici problemata,* in *Medicae Artis Principes* (Geneva, 1567), problema 68: "Why the surface of the skin is ulcerous in those suffering from fever." Incidentally, the final word in the phrase is given erroneously in the published transcription as *tentarunt* instead of *tentarit.*

44

4, Contained: morphew, worms, epidermis for bark.

5, Various color: *white and fayer, pale: darker*. Spaniard *tanned, sunburnt, tawny mores fess, Gypsy colored, Cole black Abisini Congo; Ash color east indyes, olive color west indyes; betwene green sunburnt and black*.

6, Use: means of touch, of defending the skin; detains the moistures like a blister; polished, smooth, wherefore better for cosmetics, unguents, baths &c., like the Turks. The ancients cared for the epidermis. In some animals the epidermis is fortified by a slippery shell; *dogfish;* by spines, feathers, hairs, squames, *bristles,* wool; in the woodcutter; straw; in the tortoise. Man *naked* &c., *yett Natur most sollicitous;* gave the faculty by which all things [obtainable] *scin, wooll, furres* &c.

Fat, what [it is]: site, quantity, substance and passions, 13r generation, problem.

According to Aristotle fat [is generated] from water, earth, air;[90] therefore [nonsanguineous animals do not have fat]. The best protection by which ingress [of cold is prevented]. Therefore I do not cease to marvel at women [fatter because of colder temperament than men, but more heavily clothed]; *furd mantle;* therefore the lean since [they have] little fat [because of drier temperament, are] melancholic.

Many services; like milk and sperm, useful goods &c.; salubrity for the rest of the coction of blood.

Site: because of the aforesaid services, nature located it in many places, [but] not in the head, brain &c., but in the lower venter for [the sake of] coction; under the skin; muscles; above the kidneys, but not in the kidneys; therefore sheep, cattle [must not be overfed since such may lead to complete envelopment of the kidneys by fat and cause death].[91]

In those in which the viscera small, veins slender. Bal-

---

[90] *P.A.*, 651a:30, where Aristotle actually says that fat is composed of "a small proportion of water and is chiefly composed of earth." There is no mention of air.

[91] *H.A.*, 520b:1; also *P.A.*, 672a:35: fat of the kidneys in sheep produces rot.

ance of the viscera and of the flesh. Wherefore little blood and fear of phlebotomy; for as silted channels &c.; and thus the viscera, liver, stomach, intestines small; wherefore they eat little.

What fat is: a supply of nourishment located in the venter; in a small [venter] within the flesh. Fat in the obese, in those with small veins, in the viscera of the stomach. All things become fat either in the abdomen or especially within the skin. In those in whom a small venter fat arises within the skin; on the other hand, in others the venter corresponds to the face.

In the younger in whom the flesh is rarer [fat arises] within the flesh; on the other hand, a defect [of age] *yeong geese &c. like a pudding* wherefore *ould geese dripp awaye and an ould goose* with compressed rump and no [fat]; [according to] Nicolò Massa [92] X contrary to Fallopius and WH. Balance of fat and flesh of a youth; fleshy older persons.

Quantity, supply of nourishment; Fallopius, Tuscan cows;[93] [according to] Aristotle Sicilian sheep are dispersed late [in the day to pasture to prevent overfeeding].[94]

Nicolò Massa [regarding] very unhappy poverty &c.[95] In ailments with loss of appetite [the fat] is first consumed by the flesh and thus more swiftly is there recovery;[96] rabbits in 4 days. In some the venter iij fingers, the nates, palms. *Trip* [in] *man 6 stone* when too much nourishment taken and fat in skin &c.

Harmfulness of fat when too great a quantity is present, 13v since in such cases [there is] too little flesh and blood; but

[92] Massa, f. 8ʳ: " . . . in newly born children there is no [fat]."
[93] Fallopius, *De Partibus Similaribus*, in *Opera Omnia*, I (Venice, 1606), 135: "in Tuscany the cows are killed when they are pregnant because they then have almost infinite tallow, and yet the foetuses are not affected by fat."
[94] *H.A.*, 520b:1.
[95] Massa, f. 7ᵛ: "After the removal of the skin you will see a noteworthy amount of fat, but in the newly born and in some bodies emaciated through lack of food, or for some other cause, as was seen in that very unhappy year of poverty 1527."
[96] *De Usu Partium*, IV, xi: "It is consumed from the flesh and so [the patient] is more swiftly recovered because the fat is converted into nourishment]."

46

when there is little [which is] very hot and melts away, it is an excrement. Wherefore fatty gangrene swiftly putrefies into ulcers.

Substance. Similar to *greace*, very easily melted. *Hog's* belly produces STEAR, *tallow*, sevum [=tallow]; in the heart, very bitter fat medium between [that of] swine and cattle, viz., of man. WH 3rd kind, *Buttery*, *oyly*; mingled [with] liver of fishes.

Fat: tongue, flesh of eels, *gravey*, pesti Italice [Italian dinners]. In dropsy *curded; fatt vomited or in glister cake sope*. Wherefore steatoma; leucophlegmasia *cromy and slime*. Color becomes flavus, jaundice, *yellow Hammer*, in the aged.

Generation disputed and therefore qualified. Material: oily blood, vapor, cold; from a portion of blood; cold, sticky and slimy [according to] Fernel.[97] [According to] Fallopius for the sake of something in its first formation.[98] WH: furthermore,[99] the same [effect] on semen and hair, the latter becoming swollen and loosened,[100] &c. &c. Therefore obese sterility and symptoms of lues venerea. WH I see by the veins and arteries that [the fat] in the omentum [is] from the blood, as [in] all the rest of the parts; and the fleshes are consumed by coitus.

False fat, flesh. Obese sterility, because deficiency of heat the producer of fat; from much material little as an elixir; wherefore [according to] Aristotle obesity an affection of grapevines;[101] wherefore less spermatic, as grapevines harvested in the fields.

[97] *Methodus Medendi*, in *Universa Medicina* (Paris, 1567), p. 312: "Fat . . . lines the tongue and parts of the mouth with a kind of stickiness."
[98] I.e., a necessary part, not as according to Aristotle an increment even in origin. Fallopius, *De Partibus Similaribus*, ch. 6, in *Opera Omnia*, I (Venice, 1606), 135.
[99] For this word the published transcription gives *Sar*, which properly seems to be *Super*.
[100] The Latin given in the published transcription for these last two words is *inflammantur*, *tensiva*, although properly they seem to be *inflantur*, *laxantur*. It was believed that too much fat produced baldness, as in Harvey's succeeding remarks, and according to Aristotle, caused impotence.
[101] *G.A.*, 725b:29: "Some have much semen, others little, others again none at all, not through weakness but the contrary, at any rate in some cases. This is because the nutriment is used up to form the body, as with some human beings, who being in good condition and developing much

For the efficient cause and the temperament according to some a very great heat desirable; [but according to others] a great cold; [according to] Galen. WH It is tempered; [according to] Galen oily somewhat humid; because by hot temperament; combination of water and earth.[102]

Cold proves nothing; 1st, the same regarding flesh so that they say (as some) that in the egg, wherefore not sufficiently philosophers; they suppose X for neither heat nor cold air aids; 2nd, they make generation separately as X from the veins of animals because in fever, heat and member so that of the veins again into oiliness. WH Native heat wholly the efficient cause for the variety of places as the hammer the instrument of instruments.

Employs cold as in the generation of flesh, so that *Molten lead*[103] [solidifies immediately when removed from fire]; nothing except passivity; does not generate fat; [according to] Erastus X a concretion [occurs] from cold[104] for if at the same time there were concretion and generation, the same thing would be set in motion at the same time by two opposites.

Problems of fat. Men do not grow fat.   14r

1. Disorders of the spirit wherefore: thin students [due] *to working*; fat boys somnolent.

2. They are rendered thin by too much unseasonable coitus so that all animals *crestfallen pin-Buttockt*.[105]

3. *Phisick, beere, salt, wine, vinegar &c. mustert* sugar. Wherefore [according to] Caesar the Germans ["suffer no importation of wine whatever, believing that men are

---

flesh or getting rather too fat, produce less semen and are less desirous of intercourse. Like this is what happens with those vines which 'play the goat,' that is, luxuriate wantonly through too much nutrition, for he-goats when fat are less inclined to mount the female; for which reason they thin them before breeding from them, and say that vines 'play the goat,' so calling it from the condition of goats."

[102] *De Simplicium Medicamentorum Temperamentis*, II, 20; Kühn, XI, 511–513.

[103] The published transcription incorrectly gives *Molten lerd*.

[104] *Methodus Medicinae*, in *Varia Opuscula* (Frankfurt, 1590), p. 26.

[105] I.e., pin-buttock, a narrow or sharp buttock; cf. Shakespeare, *All's Well*, II, ii, 18: "It is like a Barbers chaire that fits all buttockes, the pin buttocke, the quatch-buttocke, the brawn buttocke, or any buttocke."

thereby rendered soft and womanish for the endurance of hardship"];[106] slender diet; London boys.

4. Athletic constitution degenerates into cachochymia[107] in diseases which weaken the body. Wherefore *too pampering browne bisket in ye galleys*; maniacs. Great alteration in 1,000 years: *finer, wittyer*, &c., as in Spartan laws especially caution.

Partly membrana carnosa, partly membranous, wherefore differs in animals and parts; in dogs and horses, binds the venter; shaking off flies &c.; the elephant kills [by weight]; in parts where more fleshy the duty of muscle; wherefore I believe the ears hairy; Vesalius knew [of men who could move] the skin of the chest;[108] the fleshier *flancke* &c.

Use by which the skin retains heat, muscles; and according to some it generates fat in some by cold in others by density. Site in man under the fat; on the other hand in apes and dogs &c., wherefore in sheep it adheres to the skins. Acute sensibility wherefore rigor in catarrh is cold *as a payle of cowld water powred on Neck*. WH Is obliterated by divisions between fat, for [it can be] divided infinitely.

I consider muscle in [regard to its] general parts, particular divisions &c. Action of muscle, contraction and lateral dilatation, wherefore approach &c. Not only of the tail [of muscle] &c. For example, the upward movement and progression of the arms, body. With one at rest the other &c.; both at rest, compressed. 14v

1st, the flank at rest, constriction and dilatation of the chest; by a contrary way in respiration: adducted by the large; abducted by the small. For example: on entrance into cold water, a grimace. *Beare att ye stake pant; cockfighting*.

2nd, the flank at rest, erection of the body, tonic motion. Wherefore contrary to the common antagonist of the

---

[106] Caesar, *De Bello Gallico*, IV, 2; H. J. Edwards trans. (London, 1926), p. 183.
[107] Cacochymia means literally a depraved state of humors. The term was still in use in the eighteenth century.
[108] *Fabrica*, p. 192.

loins; special use because nature [does] nothing in vain; wherefore [observed] through the example of animals, boys, thin persons, *in tumblers*.

2nd, with the chest at rest *in tumblers*, in coitus; *leaping swinging*.

3rd, with both muscles at rest: vomiting, voiding, urine, birth, flatus; but not because of this only and especially as the common [belief]; for example, birds and animals; then the body is curved in voiding, wherefore X from the chord of the arc.

4th, by flesh to make warm, wherefore in thin persons warming [is] likewise for protection. For from fibers and the head [of muscle] aliment affects, before, above. Some ascending &c.; for thus they compress uniformly and are held fitly; *as the carior his pack: lo: cros:thuart*. Some oblique &c., so that they engage and compress the body on all sides.

The number [is] 8: 4 pairs and a fifth pyramidalis on each side; 5.

They are extended, raising the neck and scapulae which have been bent back 1, Two external oblique descending;[109] form of a very 15r large Δ. Flesh from the top of the bone of the ilium and per-forated by sinews from the os pubis; [from the] lumbar membranes;[110] [from] 7 or 8 ribs; digitated by the serratus major [and inserted] into the linea alba, which from the gathering of tendons from the mucronate cartilage[111] to the pubes [is] wider around the umbilicus. Use, dilatation of chest and inclination to the sides and compression where the muscles compress the venter.

2, Oblique ascending[112] intersects the former in the shape ⚊; fleshy from the appendix of the bone of the ilium and the spine of the os sacrum; [from] the membranes of the back and the extremity of the four lower ribs; it enfolds the recti in a wide tendon to the linea alba and os pubis; not separable from the recti according to X anatomists. These four [muscles have] veins and arteries from the vena mus-

---

[109] The external oblique muscles of the abdomen.
[110] The lumbar fascia.
[111] The xiphoid process of the sternum in modern terminology.
[112] The internal oblique muscles of the abdomen.

cula arisen near the loins; nerves from the lowest thoracic vertebrae. Uses: like the former they compress but in another place.

3, The recti arise from the anterior and superior part of the os pubis; then [becoming] fleshy, [insert] at the sides of the pectoral bone into the cartilages of the true ribs in a wide and fleshy termination. 3 sections, more rarely 4 divisions for strength and power for bearing. For when [divided into sections, usually] 1 below the umbilicus, 2 above.[113] Special use: to hold the body erect, to raise it when it is bent forward or backward by the force of one muscle, [but] curved by two. Thus its nature partly one, partly several. They are extended, raising the head which has been bent back

Sometimes four nerves at the middle to their impressions; from the extremity of the thorax. Veins called recurrent from the hypogastrium and breasts of which anastomosis, which appears rarely; harmony between uterus and breasts;[114] WH but in another way a consensus by coöperation in the effort. There is agreement also in larynx and chest and uterus.[115] Nor is the effort of regurgitation of milk from the place where it is not, but from [where] it is, viz., the veins.

4, The transverse muscles, the undermost, arise from the aforesaid membranes or from the ligaments of the vertebrae; fleshy sinews on the inside of the ends of the ribs, in the bones of the ilium; attached by a wide tendon, above to the ensiform cartilage and linea alba, below at the pubis; adhere strongly to the peritoneum; wherefore scarcely [can be completely separated from it], and what has been separated [can hardly be joined]. 15v

The tendons of these, as the oblique, are perforated for the umbilicus and for the spermatic vessels, and also in

[113] The rectus abdominis muscles are each divided into sections by transverse tendinous bands. These are placed at the xiphoid process, halfway between this and the umbilicus, and at the umbilicus. Only exceptionally is one placed below the umbilicus as Harvey suggests.

[114] A persistence of the ancient teaching that the uterus was connected to the breasts by veins. To casual inspection there is a chain of vessels that runs from the pubis to the chest (inferior epigastric, superior epigastic, internal mammary), but they do not link uterus to breast.

[115] This refers to the changes that take place at puberty.

woman for the processes of the uterus and cremasters;[116] wherefore bubonocele[117] [occurs] in them.

5, Slender [muscles called] pyramidales [approach] one to the other, larger from the pubic bone; sometimes they are [wanting], sometimes [their function performed by] part of the recti, wherefore not necessary.

Uses as of little moment and not necessary; for by nature in uses of little moment, the parts of little moment; uncertain in an obscure place; to elevate the linea alba and to compress the bladder; [also called] the succenturiali or aiders of the oblique ascending. Guin.[118]

Observation of Nicolò Massa [regarding] sensitive tendon[119] &c.; convulsions and pain; bad symptoms wherefore section of calculus. Paracentesis is performed 3 fingers below the umbilicus, 4 at the sides because of tendons, veins; consolidation in the fleshy part with little mobility.

WH From resistance by thickness Δ by callous

Peritoneum stretched around; in Arabic [is called] 16r siphac; the whole internal [lining of the venter];[120] very thin substance like a spider's web, very dense, strong and light, so that ["when the Belly is full of meat, or the Wombe of the burthen, it might be without danger stretched and relaxed as wide and long as neede required"];[121] in men to the upper venter;[122] in women above the uterus at the liver;[123] sometimes designed especially for

[116] Refers to the round ligament of the uterus. The cremaster muscles are better developed in the male.

[117] A hernia.

[118] Matthew Gwinne, M.D. (Oxon.), first professor of physic at Gresham College, with a flourishing practice in London until his death in October, 1627. He was the author of two Latin plays, but left no medical writings of any note. See Munk's Roll, I, 118–121.

[119] Massa, ff. 12ᵛ-13ʳ: "You will note also that the didymi, which are of the substance of the peritoneum running to the scrotum, perforate those tendons of the muscles, whereby in the descent of the intestines, or of the omentum, there often occurs a division of the tendon; therefore let those who attempt to cure such patients be careful lest by violent treatment the tendon be lacerated, since terrible results arise from this so that there are pains, spasm and even death."

[120] I.e., inferior cavity encompassed by it (Crooke, p. 78).

[121] Crooke, p. 78, citing Galen, De Usu Partium, IV, ix.

[122] In men it is thicker from the ensiform cartilage to the navel that it not be a burden to the parts beneath (Crooke, p. 78).

[123] In women . . . from the pubic bone to the navel for the sake of pregnancy (ibid.).

52

the stomach; os sacrum. Double wherefore all things contained, stretched around on all sides. It is said to arise at the back. WH in none does it arise or terminate; everywhere firmly [attached] by muscles; in the back [it is thicker]; all parts are contained in duplication, of the second rule; WH because they provide a little tunic for all things.

It is perforated for the oesophagus, vein [and] artery,[124] sixth pair of nerves,[125] umbilicus, anus, bladder, uterus &c. In the part where it is perforated it appears double &c.; point d'ore; subject to rupture, scar. These extensions opened in mice; uses for all; of the liver &c.; it provides a covering, contains and strengthens; it separates the intestines from the interstices of the muscles lest rupture occur; WH I have here one ruptured; use of Plater.[126] A kind of torcular sac.[127] Dr. And:[128] medium weight, medium □;

Vitruvius ⨉ ; woman sexual pleasure through umbilicus.

Umbilical vessels, veins, arteries. Here is not the place but in [a consideration of] the embryo; and cuticular separation in the air [i.e., after birth];[129] [according to] Aristotle 4:672,[130] umbilicus wide in man *like a Button*

[124] Refers to the inferior vena cava and aorta which lie behind the peritoneum.

[125] The sixth pair of nerves originally included the modern glossopharyngeal, vagus, and accessory nerves. Here Harvey refers to the vagus nerve.

[126] Plater, p. 148: "The general membrane which surrounds the venter and is called the peritoneum, from which the remaining members advance. It contains and constricts the viscera and vessels lest they become involved with the muscles or break forth under the skin and become ruptured; nonetheless since it can be dilated, when it is distended it yields sufficiently for the stomach, intestines, bladder and uterus."

[127] *De Usu Partium*, IV, i; cited by Crooke, p. 80: as when a man squeezes a bag of herbs from every side, he may more equitably strain the liquid out from all sides.

[128] Probably Richard Andrews, M.D., several times Censor of the Royal College of Physicians. On April 25, 1631, he was named as Physician to St. Bartholomew's Hospital after Harvey's resignation or death, but Andrews never lived to take office (d. July 25, 1634). *See* Munk's *Roll*, I, 154–155.

[129] Harvey refers here to the separation of the umbilical cord, with its contained vessels after birth.

[130] The particular edition of Aristotle employed by Harvey has not been traced, but his references occasionally indicate that it was one in which the fourth volume contained the biological works.

suspended liver and blather;[131] vessels and ligaments having been made;[132] lying on the back or erect which use not in beasts.

Umbilical vein in the hollow of the liver within little mounds; I know somewhat the hollow through the same space. Two arteries in the os sacrum;[133] *sumtimes not devided but after his enterance.* Section of the umbilicus in an old woman showing the nature. **WH** URAGOS seems to be lacking in man; I did not find; **WH** it is present in brutes.

Inferior venter

Parts contained: umbilical arteries, urachos, umbilical vein, omentum, liver, stomach, gullet, mouth of stomach, spleen, intestines: duodenum, ecphisima,[134] jejunum, ileon, colon, caecum, rectum; kidneys, ureters, bladder; vessels: preparantia,[135] deferentia, uterus. Mesentery, mesenteric veins, pancreas (porta coeliacae, branch to liver; duct of the gall bladder, nerve to liver here joined), gall bladder. Aorta and descending emulgent veins.[136]

Briefly I wish to treat the site and position of all these as I know them, for you afterward [to treat] individually of each.

The site of all partly certain, partly uncertain; nature *Romidges as she can best stow; as in ships* for the sake of ease of movement; [in a case of] abscess[137] of the liver, *the gutts thrust att one side and ij fingers beneth the navill, full or empty the colick gutt on ye liver beneth the navil sitting or*

[131] *G.A.*, 740a:25 ff.
[132] The obliterated umbilical vessels on the inside of the anterior abdominal wall remain as ligaments (falciform ligament, median umbilical ligament, lateral umbilical ligaments).
[133] Here Harvey refers to the obliterated umbilical arteries.
[134] The term ecphysis was used for a process or appendix, and, by Galen, for the duodenum. The duodenum was also called *dodecadactylon ecphysis* (Motherby, *A New Medical Dictionary*, 1791).
[135] The old name for the spermatic cord was *vasa preparantia*, for it was believed that the seed was made there. The term *deferentia*, or in medieval anatomy *delatoria*, was used for the epididymis.
[136] The renal veins.
[137] The published transcription erroneously gives *imposterum*, which appears to be and makes sense only as *impostium* or impostume.

54

standing as opposed to *lying*. In coughing,[138] *breathing* [the intestines] arc moved; **WH** [and also] Δ by many things, never preserving exactly the same position; [in the] pregnant [woman]; *yeoung girls by lacing;* wherefore *cutt there laces.* The ilia suspended: Cardinal Campeggi; *hard and yet* to pulsate; hypogastrium, *cleane empty*.[139] The intestines sometimes inflated beneath; sometimes contents retracted, a bad sign of weakness of intestines.[140]

The whole liver more on the right,[141] and the umbilical 17r vein[142] [is connected to it]; see the connection to the spleen. **WH** My swelling in a quartan;[143] *under the Choin* for protection *allong vp to the 7 Ribb* on the right above; wherefore difficulty of respiration by reason of the swelling

[138] This word, *tussiens* in the manuscript, is not included in the published transcription.

[139] Colombo, p. 492: "In the case of Cardinal Campeggi all the intestines had been elevated to the hypochondria and therefore the lower abdominal cavity was empty of intestines." The cardinal must have had an enormous ventral hernia, possibly umbilical.

[140] Crooke, p. 166: " . . . when the circular Fibers do gather themselues from below vpward . . . and in this motion the wind, the Chylus, & the excrements are auoyded by the vpper parts, nothing by the lower although such euacuations be prouoked by sharpe Clisters. . . .

" . . . Inflammations streighten the passages, oppilation altogether shutteth them. So then the excrements hauing no currant passage, the expelling faculty following first of all the order instituted by nature, beginneth her contraction in the vpper part, to expell the excrement downward; and this it endeuoureth againe and againe, but being frustrated because of the streightnes or stoppage which opposeth her; she inuerteth the order, and begins her contraction from below, and that with such violence, that (alas the while) the Chylus and the excrements are thrown out by the mouth: so diligent and circumspect is Nature to vnburden it selfe of that which is noysome or offensiue.

"Againe, a light exulceration may cause this depraued motion; as, when the gut is afflicted by the proritation or goading of the vlcer it transmitteth vpward such things as would offend it, & as it were altering her saile, beares her course vpward against nature, which before she held downward. This *Peristaltick* motion which is contrary to the naturall, those haue miserable experience of which are afflicted with the lamentable and odious disease called *Ileos* or *Miserere mei Deus*, wherein the seate or fundament is so closed, that a Needle cannot be thrust into it, & if any Clisters be with much ado administered, they are incontinently sukt vp, the circular Fibres contracting themselues from below vpwards." Cf. nn. 243, 256.

[141] The right lobe of the liver is much larger than the left.

[142] The umbilical vein here refers to the ligamentum teres.

[143] It is clear that Harvey had suffered from malaria, as did his sister, cf. f. 36ʳ. The association of splenomegaly with malaria was noted by Hippocrates and is mentioned in the ancient medical writings of India and China.

of the liver *long the short ribbs vpon the stomach which it covereth.*

Connection always to the diaphragm by two very strong ligaments;[144] to umbilical branch of vena cava;[145] to the spine; sometimes to the ribs by peritoneum, to colon. It is connected to the head through the nerves of the heart and vessels, to the stomach and spleen through the splenetic branch.

Splen other side of the stomach towards the short ribbs. It is touched by the hand *vnder the short Ribbs att the end of the* last *or* penultimate; *soe vnder and soe behinde* so that it is scarcely felt in the healthy, especially when the belly is tense or fat. When it is swollen nothing is more easily felt by touch, and it descends.

It is connected to the omentum; sometimes in animals is seen to be connected to the diaphragm, to the peritoneum, to the left kidney. With the spleen in place of the liver and contrary to Aristotle;[146] *is mayde and scente* rather to the right if the white[147] part [is] anterior; with the liver, in the middle, in three lobes. *So* [in the case of] *Rowena Chirn.*[148] A pendent[149] letter V will be on the right, in the anterior and inferior part.

Omentum above the intestines, therefore [it is called] 17v epiploon. Is not in all sanguineous [animals]; in fish, seal; not in all quadrupeds, *ratt and mows;* not in all fowl, as the pigeon &c. Partly like the pia mater of the brain; twistings of the intestines; shape, in two membranes *powch reversed*[150] in the upper part of the stomach *along thother Colick.* Below *sum time,* and at times very small, *retracted* between stomach and spleen wherefore rarely below the

---

[144] The right and left lateral ligaments.

[145] The ligamentum venosum of modern anatomy.

[146] *P.A.*, 670a:1: "It is the position of the liver on the right side of the body that is the main cause for the formation of the spleen; the existence of which thus becomes to a certain extent a matter of necessity in all animals, though not of very stringent necessity."

[147] The published transcription gives *alta* rather than *alba.*

[148] The published transcription gives the meaningless "Sorow Thom:." For another reference to Mrs. Chim, cf. f. 46ᵛ.

[149] The published transcription gives *prudens* instead of *pendens.*

[150] This refers to the lesser sac of the peritoneum.

umbilicus in man; very often it is convoluted between the Omentum turned back intestines and spleen. See *fayrely spread* but the colon in the same place below the umbilicus, extended beneath; X always ignorant anatomists [believe] it descends to the vagina of the uterus; wherefore sterility WH X. Epiplocele: Colombo Book I.[161]

Connection: above, one part to the fundus of the stomach, sometimes to the liver;[162] the other part to the colon (but not in the dog) wherefore for the sake of the mesentery and colon; wherefore its position is more uncertain. *Below loos sometimes to the* os pecten[163] but rarely. Before, Omentum expanded behind. Sometimes they adhere to each other[164] and to the peritoneum. Here and there sometimes to the sides, always to the spleen; it arises from the second vertebra from the peritoneum *where* the arteries and the portal vein *going* to the liver *ar slightly tyed to* the jejunum.

Passions of the omentum; [according to] Apollonius filled with black humor;[165] according to Riolan swellings of the hypochondria and around the vulva [arise] from humors collected in the omentum.[166]

Receptacle of filth of the spleen Δ. [According to] A[cqua]pe[n]d[ente] filled with much black humor;[167] belly-talk from flatus of its pouch dispersed to the glandule.[168]

Sometimes the quantity is very slight so that it can not 18r extend to the umbilicus. Swiftly consumed in fasting. Wherefore some believe [it is] to nourish; WH X but be- Expanded

[161] Harvey's reference is obscure. The reference is correctly to bk. XI, not to bk. I as in the published transcription. In the edition of 1590 this would be pp. 422–425, which deal with the colon, mesentery, and omentum.
[162] The lesser omentum is placed between the liver and the lesser curvature of the stomach.
[163] The pubis.
[164] The layers of the greater omentum frequently become fused together.
[165] Hippocrates, *Epidemics*, bk. 3, 13th patient; Littré, III, 137 ff.
[166] Riolan, p. 166.
[167] Hieronymus Fabricius, *De Omento*, in *Opera Omnia Anatomica et Physiologica* (Leipzig, 1687), pp. 123–128.
[168] Du Laurens, p. 225: "Finally I add from Hippocrates' book *De Glandulis*, that a copious humor flowing in from the intestines, when it cannot at once be received and absorbed by the glands, is preserved in the omentum as in an estuary."

cause more liquid than flesh it is swiftly corrupted, as [observed] in cadavers; and more swiftly corrupted because all fat; wherefore according to Hippocrates having come forth from the body necessarily it putrefies;[159] wherefore [according to] Galen in the case of a wound[160] it ought not be replaced.[161]

Proportionately a large quantity in man, wherefore epiplocenasti,[162] omentum-carriers. Among animals more in the dog and ape; tripe [in] *man 8 stone more then an oxe*. The larger in which the fat separated, wherefore in the older; *yeoung ox less tallow*. Because it becomes fat; *carene*[163] within *it furreth vp the veynes* as silt in a canal.

Substance of the membranes thin, light.

Division and reflection
3 divisions: [1] *fatt grese* in which adeps, *tallow* in which sevum; [2] ample veins, a sign that the fat [is] bloodless, because that blood not very oily; [3] glands for distribution of the veins.

3 uses: [1] it carries down veins from the portal vein to the stomach and spleen and to the colon and duodenum; [2] all things collect [there] instead of [in] the liver, stomach &c. &c.; [3] and it warms all the back, lubricates the intestines for their motion and coction.[164]

Fallopius
None
Because (Avicenna [says]) the concocting heat receives and retains the stickiness; wherefore [according to] Galen a *fencer* [wounded by a thrust to the abdomen so that part

[159] *Aphorisms*, VI, 58: "If the omentum comes forth, necessarily it putrefies."
[160] The published transcription erroneously gives *vulcer* instead of *vulnere*.
[161] *De Usu Partium*, IV, ix.
[162] More correctly this should be *epiploocomistes*. This term was applied to those with a large omentum and possibly as a term of raillery for those with a hernia containing omentum—particularly an umbilical hernia (Motherby, *op. cit.*). Crooke, p. 99: "Of men, those that liue a sedentarie and idle life, sacrificing to their appetites, haue it so great, that it becommeth a burden unto them, whereupon they are called *Epiploocomistae*, that is, Kal-carriers."
[163] *Carene*, a now obsolete word meaning a thick, sweet wine.
[164] *P.A.*, 677b:30: "The development, then, of the omentum is simply the result of necessity. But when once formed, it is used by nature for an end, namely, to facilitate and to hasten the concoction of food. For all that is hot aids concoction; and fat is hot, and the omentum is fat." The published transcription has as the final words of Harvey's sentence *unde inter super. et coctioni*, which seem rather to be *ad ipsarum motum et coctionem* and have been so translated.

of the omentum fell out];[165] without a wool [covering for
the part suffered from] cold; wherefore fat and jovial persons eat everything; the emaciated without food and the ill
eat a lighter diet.

Stomach: above [is] the mouth of the stomach; a little to
the left [below] the diaphragm. Below [are] the omentum
and colon, and sometimes above; wherefore nausea from a
clyster; and on the other hand flatus downward. In front it
is touched by the epigastric region and *soe along the short
ribbs and soe in* to the left where [is] the larger part and
closer to the spine, place the pancreas and portal [vein]; at
the right the liver, and at the left the spleen.

18v

Therefore
when faeces in
the colon the
appetite is lost
Some [have]
coction from
faeces as
fermentation
lowest *chimie;*
Cl[yster]

It arises from the oesophagus and it ends in the pylorus
(pyle, door, doorkeeper) a little below. That it may give
place to the liver, it approaches the gall bladder; wherefore
in some the ardor of the stomach crosses through the coeliac [artery] to the porta [hepatis], to wit, the portal
[vein] particularly of the stomach.

Veins from the splenic branch of the portal because [1]
they carry blood; [2] they carry down chyle; therefore it
is not nourished by chyle according to Galen III [7] On
the Natural Faculties;[166] wherefore it swiftly nourishes its
strength.

Coronary vein[167] and short vessel; from splenic after divisions in the superior part; wherefore it regurgitates a
hunger-provoking acid. WH X it does not differ from the
rest of the veins. Cardinal Cibo; [according to] Colombo
it was dilated, vomited blood.[168] Regarding many things,
[according to] Riolan and Δ WH varices [discussed] de-
lightfully in On the Humors ch. 29. Arteries from coeliac
which [will be discussed] elsewhere. Two nerves from

[165] *De Usu Partium*, IV, ix.
[166] Galen, *De Facultatibus Naturalibus*, III, vii; A. J. Brock trans.
(London, 1928), pp. 249 ff.
[167] Now called the left gastric vein.
[168] Colombo, p. 492: "in Cardinal Cibo a vein which extended from the
spleen to the mouth of the stomach was of notable size; when this was first
opened because of anastomosis or rupture, immediately the stomach was
filled with blood, and not only the stomach but all the intestines. It is possible that while he was as yet living, in his sickness he vomited up several
pounds and then declined through its loss."

which the recurrents come, which crossing through the diaphragm are divided into two, right into left and similarly the left at the right, and at the posterior of the mouth of the stomach and then at the pylorus; wherefore the mouth of the stomach [appears] almost wholly of nerves; wherefore the appetite [is] in this seat; two other nerves from the 6th pair from the roots of the ribs.[169] Sometimes nervules to the left side from those which [seek] the spleen. Wherefore a great consensus in the cerebrum and, on the other hand, therefore vomiting and headache in the injured; wherefore with the cerebrum affected weakness and melancholy are suffered.

Hence astonishing hope and trust, and on the contrary, 19r by affections of the mouth of the stomach; wherefore *Nerve stick vntill att hart*, wherefore excitement of animal by function of the veins; *fatness; motion from the lower Hous;* thus acid humors, acid *sawces;* wherefore if too much heartburn, pains, fainting, vertigo, epilepsy, &c.; wherefore in me myself nausea and sneezing; wherefore fever little increased in epilepsy; wherefore hiccuping from absorbed or compressed humor; wherefore *Sir William Rigden all yelleow;*[170] wherefore *J. simpson of Chalis,* detention of spirit; wherefore Sal[amon] Albert[i] from compression of the ribs.[171] **WH** Sneezing; wherefore *in a frogg* only the throat touched; wherefore by valetudinarian diet of lessened sharpness; many are always complaining; *Lady Croft;*[172] in embryos mucous, air; lest there be hunger.

The chest having been pierced and the spleen dilated

---

[169] This would seem to refer to the splanchnic nerves or the sympathetic chain. There was much confusion at this time between the vagus and the abdominal sympathetics.

[170] Sir William Rigdon, of Dowsby Lincs., was born in 1558 and knighted on July 23, 1603. His will, dated October 25, 1610, was proved on November 20, 1610, so that he must have died between those dates. His executor and heir was his son Robert, who entered Cambridge in 1605, married Mary, the daughter of John Argent, and died May 12, 1657. See *Lincolnshire Pedigrees,* ed. A. R. Maddison (London: Harleian Society, 1904), 52, pp. 820–821.

[171] "Thema Medica: de singultu," in *Tres Orationes* (Nuremberg, 1595), f. [59ʳ▬] Hˢʳ.

[172] Wife of Sir James Croft, Kt., of St. Osyth, Essex, a pensioner of Queen Elizabeth's bodyguard and grandson of the Controller of the

Intestines arise from the stomach and end in the anus. 1 *smale gutts* then *greate; great gutts* circumvallating everything *as horse*. Duodenum from pylorus under the stomach placed under the exit of the mesentery behind the connection to the back by many ligaments and the pancreas ends under the colon.

Twelve fingers in length which is not [found] in these times.[173] Shape not convoluted as the rest because the liver is very near; that the descent [of the portal vein] be not impeded. Jejunum because [found empty]; from the many veins [drawing off] the liquid chyle; dense bile enters from the end of the duodenum [and passes] through 12 palms and iij fingers, measure of 5 feet; wherefore it is there changed in color and there are not so many veins nor so great emptiness.

The ileon [is] circumvoluted wherefore the wretched 19v disease [called ileos],[174] when [excrement and vapors return from the colon into] the mouth of the ileon. It arises where the intestine begins to redden and ends in the colon under the right kidney; it occupies the whole lower region under the umbilicus and colon. Above the ilium and its crest, the bladder &c.; [it is] 21 feet in length; [according to] Nicolò Massa 20;[175] caliber less than colon; *smale gutts*. Shape in man, in every one convoluted *vnsett Ruff*; on the other hand, [in] *Birds and fish*, pigeon, goose, *Hen* [straighter]; *Barble playse; partly vp and downe partly* here and there.

Because in man the intestine [is] 6 times the length of the body, 7 times [in the] *ginney Cunney*, but 6 times in man and many large animals. That the animal need not al-

---

Queen's Household. From 1619 onward, James Howell, who was also a patient of Harvey's, addressed to him some of his *Familiar Letters*. He is mentioned in Lord Herbert of Cherbury's *Autobiography*.

[173] Here Harvey suggests the idea that was common in the sixteenth century, to the effect that the failure of anatomical findings in dissection to agree with the statements of Galen resulted from man's physical degeneracy during the centuries following the era of Galen.

[174] Cf. n. 140.

[175] Massa, f. 21ᵛ: "Its quantity is not the same in all; I have found twenty feet, but very often less."

ways feed, therefore more convolutions in them; the very voracious fish [has a straight gut].[176] Large animals retain aliment as plants, because it is not necessary for them to eat as frequently and to excrete [constantly]; by the example of the candle. Wherefore elephants [have] almost 4 bowels;[177] wherefore Δ 12 times [its length] in the porpoise which has no large intestine; wherefore smooth and accordingly lacking in size; wherefore long intestine of small [caliber]; on the contrary, large as stomach [is] shorter; wherefore the stomach [is] either thick, [or if] several [of them], smaller; on the other hand, [if] only one stomach, larger.

6 times likewise in man and in many large animals because they concoct more slowly, because their food is more completely elaborated; in man so that more time to employ in the other duties of life.

Shorter because rougher nourishment because more time to void; ravenous feeders, they concoct easily and swiftly. [According to] Hippocrates rapid discharge.

Caecum, monoculus,[178] appendix at extremity of the 2or colon, because blind in duty, wormlike in size. In man, among large [animals, it is small] yet worthy of note. Like a caruncle; on the other hand, in *Hoggs Hare oxen Ratts* &c., like another stomach. In man sometimes large, as it is in the foetus WH; [according to] Sal[amon] Albert[i] sometimes not completely [formed].[179] Here the rule of Socrates according to similitude *in a great print*. WH transit of these things [will be considered] afterward; here only the site.

Colon, of a glutton,[180] *Gutt*. It arises from the caecum upward to the loins, crosses over the right kidney to which it is connected by fibers. It crosses under the liver and gall

[176] *P.A.*, 675a:20.
[177] *H.A.*, 507b:35.
[178] The medieval name for caecum.
[179] "Schema rerum et venarum emulgentium," in *Tres Orationes* (Nuremberg, 1595), f. [87ʳ=L⁷ʳ] : "the present cadaver displayed to the observers a defective caecum intestine."
[180] This represents a lacuna in the transcription. Harvey's word appears to be *ganensis*, although this may be the fault of his hand rather than his Latinity. Correctly it would be *ganonis*.

62

bladder, to which it is sometimes connected, wherefore I have seen it tinted with [yellow] bile;[181] whence by way of the bottom of the stomach on the left, sometimes iij or 4 [fingers] below the umbilicus. At the left upward to the left kidney to which it is connected, wherefore its relationship to the kidneys; [hence] nephritis [confused with] colic [pain]. It crosses the spleen wherefore here [occurs] rumbling of flatus; and here because of its twistings, pain [occurs]; its place of origin downward at the flanks where it is joined to the peritoneum,[182] and it ends continuous with the rectum.

Length 7 palms, 7 fingers; [according to] Massa 10 feet.[183] The colon appears here upward that food [may not press] downward by its weight; for example, *lead Bullet*;[184] *drink standing* [on] head, *Rise*; [clyster administered] by physician for colic, into the hollow [of the colon].[185]

Rectum finally. It arises from top of the sacral bone and ends in the anus. At the sacral bone it is connected to the peritoneum.[186] Under the bladder, under the uterus, under the prostate and root of the penis; wherefore agreement with bladder[187] and prostate; wherefore **WH** pain from excrement inflames prostate; wherefore [according to]

[181] Post-mortem staining of the hepatic flexure of the colon is common.
[182] This refers to the pelvic meso-colon.
[183] Massa, f. 20ᵛ: "the quantity of that intestine is not the same in all, for sometimes it has been found ten feet [in length] and often less."
[184] Cf. Shakespeare, I *Henry IV*, V, iii, 36: "God keep lead out of me! I need no more weight than mine own bowels." Cf. f. 26ᵛ. For a further possible Shakespearean reference, see f. 22ʳ.
[185] *Methodus Medendi*, XIII, 17: the substance of a clyster, according to Galen, cannot pass upward from the colon into the ileon, although medicinal force may go higher. However, Du Laurens writes: "Let the gut be dried and blown up a little, and pour water into the duodenum and it will presently issue at the rectum, but on the contrary if it be poured into the rectum it will stay in the appendix of the caecum because it can get no farther; which proves that in the end of the caecum there is a valve . . . to prevent reflux."
[186] The rectum is posterior to the peritoneum in the hollow of the sacrum.
[187] Crooke, p. 109; "Hence comes in men that notable sympathy of this gut with the bladder, which maketh a man that is troubled with the stone in the bladder, to be euery moment prouoked to the stoole; but yet in women there is a greater simpathy betweene this gut and the wombe; so that oftentimes by compassion the gutte is vlcerated, and the excrements are auoyded by the secret parts."

Galen inflames intestines [and produces] ischuria.[188]
Wherefore I have recognized and cured ulcers of the 20v
prostate and penis in the anus; wherefore ulcers from the
cervix of the uterus often into the anus,[189] wherefore
thence excrements [are voided];[190] [in some] most
wretched women I have known hard excrement descended
of the uterus; on the other hand, twisting of the uterus by
impediment of a clyster; wherefore calculus of the bladder
is touched. For example: *cutting on the finger*.[191]
All the intestines *but one differ:* in substance, site, shape,
quantity, duty, name. Substance: thick, thin, jejunum,
ileon &c. Site: above, a property of the stomach and liver
on the route of nourishment; below; colon, rectum. Shape:
straight, duodenum, rectum; twisted, jejunum, ileon.
Quantity: larger, large, smaller, small. Duty: for chyle,
slender: Δ X convoluted: hernia, Δ WH; excrementa: colon,
rectum—in animals, caecum. *from Powles to leden hale one
way but many names as cheape powtry*[192] &c. In the mesentery and all intestines at the back: *Chaudron of a calf.* Because intertangle because *fall downe on a heape;* wherefore
perhaps hernia; not of this but [according to] Avicenna of
the caecum;[193] and WH part of colon. *Figure like the stock of*

[188] A retention of urine. *De Locis Affectis*, VI, 4; Kühn, VIII, 403;
Daremberg, II, 679.
[189] Here Harvey means a rectovaginal fistula.
[190] Cf. Benivieni, *De Abditis nonnullis ac mirandis morborum et sanationum
causis*, tr. Charles Singer (Springfield, Ill., 1954), p. 165: "I have also seen
a girl who was born with the passage of the anus closed up and who every
eighth day discharged excrement from the naturalia."
[191] In the operation of perineal lithotomy the operator had an index
finger in the anus which hooked forward the calculus. Incision was then
made in the perineum and the calculus removed.
[192] The road in the city of London from St. Paul's to Leadenhall Market;
parts of the route have different names (Poultry and Cheapside) but it is a
through road. This is reminiscent of Vesalius's somewhat similar method
of describing the intestinal tract: "although the same continuous body, yet
since it extends in many different twists and turns, varies in shape. . . . Just
as in Paris the thoroughfare from the gate of Saint Jacques to that one which
is called Saint Martin, even if it is straight and does not proceed with various turns, nonetheless, if my recollection is correct, changes its name six
times, first borrowing its name from Saint Jacques, then from a little bridge,
thereafter from the church of Notre Dame, not much later from its bridge,
then from a tavern and finally from Saint Martin's church" (*Fabrica*,
p. 424). Cf. the further use of London's geography by Harvey on f. 49ᵛ.
[193] Avicenna, *Canon*, Lib. III, fen xxii, tract. 1, cap. 2: "[Hernia results]

64

a ruff;[194] *frensh mallow lef;* site: it arises at 2nd lumbar vertebra from peritoneum, wherefore agreement of back and of intestines; it crosses in all the intestines continued to the rectum.

It is composed of internal membranes; to meseraic [mesenteric] veins, very slender arteries, slender nerves; fat, as the mud of a channel; glands for distribution of the vein.

Riolan [observed that] the mesentery [had] very keen sensitivity in many kinds of ailments.[195]

The mesenteric veins [run] like the roots of a tree to the intestines; and by individual vein and artery to the liver; wherefore a question whether the arteries carry down the chyle to the small intestines; one joined to the other [until] finally [opening] into the portal [vein]. Wherefore here the portal [vein] is the trunk of the tree and enters the liver; roots [go to] the 5 lobes[196] [from] the hollow of the liver; mesenteric branches [course] downward.

All the arteries similarly run into the coeliac; part of the mesenteric, above to the left and right; below. Trunk obliquely downward under the duodenum between the stomach and intestines to the vertebrae or the divisions: cystic, gastric.

WH See more exact division between veins: the higher splenic branch similarly partly [goes to] the epiploic intestine [to] the right, behind; partly [to] the syngastric [by] lesser and larger [branches] WH; stomachica [branches are] both left and gastroepiploic. Because [all] above the mesentery; to spleen, stomach; epiploic to colon.

---

either because of assistance from a blow, or from motion, or fall, or by retention of sperm and prevention of its discharge; or by woman lying on man, or by the effort of breathing during coitus, and especially following a full meal, and similarly by coitus after nauseating satiety and a collection of flatus and a vacancy in the venter, wherefore there descends either the omentum or covering and the intestines, and especially the caecum since it is untied and loose."

[194] This description aptly fits the arrangement of the mesentery of the small bowel. When the small bowel is dissected off the mesentery the raw edge of the latter does look like the folds of a ruff.

[195] Riolan, p. 182: "I have observed many times in ailments that the mesentery is possessed of exquisite sensitivity."

[196] Harvey still retains the Galenic teaching of the five-lobed liver.

In the lower mesentery: before division; gastroepiploic [to] the right; intestinal, in duodenum, jejunum; sometimes from a branch [which] divides into 3 branches and afterward into others. Right into jejunum, ileon, caecum, part of colon; left at rectum wherefore internal haemorrhoids.[197] In such almost the distribution; arteries which [will be considered] in their place when more exact [discussion] of the veins.

Double origin of the mesentery. Above at first lumbar vertebra where arteries [coeliac and superior mesenteric] arise; above emulgents and around those which form the plexus of nerves which is disseminated with the arteries. Below at third lumbar vertebra where the [inferior] mesenteric arteries [arise].

Gall bladder [lies in] a hollow of the liver, a sinus carved out. Hence a little at the right of the umbilical vein and artery outside the lobe of the liver, wherefore in jaundice tense Δ and inflammation; *Sir James Crosby*;[198] afterward [I shall deal] more exactly with the gall bladder.

Pancreas, *sweet bread*, [is] under the duodenum at origin of the omentum, at first lumbar vertebra [and lies across] the kidneys; fat in the obese; *redder in a dogg*. Substance lax and formless; wherefore absorbing humidity [according to] Piccolomini;[199] the seat of intermittent fever. WH because worst of all [if] it is dispersed. In weakness as all glands.

Use: it sustains the veins and extends a bolster for the 21v stomach; X in animals.

NB NB Query whether the emulgents and the great vein and artery [and the vessels to] the testicles can be discussed with [those of] the rest of the intestines. If not to be inserted here with the viscera and with the pancreas, [then with] the root of the coeliac arteries and branches; likewise the artery which is disseminated in the mesentery,

---

[197] The rectum is, of course, supplied by the interior mesenteric artery.
[198] No trace has been found of a Sir James Crosby, and, indeed, it appears that the published transcription may erroneously have provided such a name rather than the more likely Crosse or Crosier. Nevertheless, nothing has been discovered about a knight of either name at that time.
[199] Piccolomini, *Anatomiae Prelectiones* (Rome, 1586), p. 100.

likewise 2 nerves: 1, of the liver, gall bladder, pylorus, 2, of the spleen, fundus of the stomach and omentum.

The kidneys [lie] under the stomach, liver and spleen, thence to the loins, under the extremity of the ribs, thence *att the sides of the great veyne*[200] as some by the folding of the peritoneum: before, behind. Wherefore some partly containing, partly contained, wherefore iij venters [according to] Diocles: head, thorax and bladder;[201] so of Fernel.[202]

Site uncertain *lower more or little over then the lower Ribbs;* sometimes on the right higher but rarely; twice only [according to] Nicolò Massa.[203] Left higher nowadays. For [according to] Aristotle and the ancients [that] on the right [higher];[204] but that is true in brutes not in man;[205] perhaps [true] in those days,[206] one now rarely [observes it]. But when the liver [is] small or on one side as in brutes, then reasonably and deservedly [the right kidney may be higher than the left], because the right is more worthy of a higher place. Site in the middle[207] when one, as [according to] Bauhin by division of the arteries.[208] Never

---

[200] The inferior vena cava.

[201] The reference is to "Diocles epistola praeservativa" in Paul of Aegineta, *De Re Medica*, bk. 1, ch. 100, in *Medicae Artis Principes post Hippocratem et Galenum* (Geneva, 1567), col. 375, which actually reads, "We divide the human body into four parts: the head, chest, belly and bladder."

[202] Fernel, *De Partium Corporis Humani Descriptione*, in *Universia Medicina*, (Paris, 1567), p. 15.

[203] Massa, f. 32ʳ: "Anatomists say that they have often found the site of the kidneys reversed, that is, the left higher and the right lower, which thing I have hitherto seen twice."

[204] *P.A.*, 671b:28: "In all animals that have kidneys, that on the right is placed higher than that on the left."

[205] Harvey is correct when he suggests that in man the left kidney is higher than the right.

[206] A reference to the idea that Galen's anatomical descriptions were in error only because of the physical degeneration of mankind since the time of Galen.

[207] Probably a reference to the horseshoe kidney first described by Berengario da Carpi, *Isagogae Breves* (Bologna, 1522), f. 17ᵛ, and next by Andres de Laguna, *Anatomica Methodus* (Paris, 1535), f. 28ᵛ. When a horseshoe kidney is present it lies just above the bifurcation of the aorta.

[208] Bauhin, p. 155: "If there should be one (which is rare), necessarily because of the required attractive force of the watery substance, it ought to equal both in size; but thus the body would not be in equilibrium unless it were placed in the middle of the spine, resting on the arched surface of the cava and aorta."

opposed;[209] *least hinder one the other* WH X. Because of inferior order, they yield place to the liver [as the more worthy organ].

Because as in the macrocosm *the weakest to the wall.*[210] Indeed, they thus give way as the inferior in the commonwealth. Thus Fernel;[211] so the ribs [are] placed apart for the movement of the heart. So *the obdurat scull* for the veins; to the rear *they leye vpon* the PSOAS muscles; they flex the leg; wherefore they are thus believed to render the legs of nephritics inactive.[212]

WH X I believe more in the agreement of the kidneys with the spinal medulla. They are connected to all parts by the little membranes of the peritoneum and omentum; to the cerebrum, heart and liver through the vessels and to the bladder through the nerves.

The ureters are connected to the bladder; above the aforesaid [psoas] muscles they cross the sacral bone between the tunics of the peritoneum, and thence downward into the neck of the bladder.

Bladder   In the cavity of the sacral bone and pubis and of each hipbone, wherefore it has a better exit in a place above the rectum intestine; in woman above the neck of the uterus, between the tunics of the peritoneum, it is connected to the pubic bone, and [beneath] the peritoneum [to] the umbilicus from the fundus through the urachos. A little higher

[209] Crooke, p. 140: "They often stand not one opposite to the other lest in their ioynt strife they should hinder one the others attraction, as Galen hath conceiued; but we, sayeth Bauhine, imagine that the cause of this position is rather attributed to the arising of the vessels and properly of the emulgent or sucking veins, because their attraction is greater and of more vse."

[210] Cf. Shakespeare, *Romeo and Juliet*, I, i, 15: "for the weakest goes to the wall." Cf. ff. 20ʳ and 26ᵛ for further possible references to Shakespeare.

[211] *De Partium Corporis Humani Descriptione*, in *Universa Medicina* (Paris, 1567), p. 15.

[212] Crooke, p. 140: "[The kidneys] lie with their flat sides vpon the muscles of the loyns which they call PSOAS, appointed for the bending of the leg; about their heads not much lower than the lowest ribs, in those voyde spaces which are betwixt the rootes of the ribs and the hip-bones. They lie betweene the two membranes of the Peritoneum, one of which lyeth vnder them, the other vpon them; whence it is that in fits of the stone, the leg on that side where the stone lieth is benumned, sayeth Hippocrates, because of the compression as well of the muscle we spake of before, as of a sinew which descendeth that way."

than the pubic bones, especially when inflated or filled. Therefore the muscles called pyramidales compress ["a part of the groine and the bladder there-vnder"],[213] because when compressed here excretion of the urine occurs. Therefore some [say] that there can be operation for a calculus here.[214] Because WH X by these three reasons.

1, because when the urine has flowed out the bladder 22v subsides.

2, when the fundus of the bladder has been wounded it does not consolidate, as Aristotle noted,[215] because [it is] membranous and sinewy; therefore it consolidates in the neck which [is] fleshy, and the more fleshy the swifter. Therefore [it easily consolidates] in children and women, [but not] in men because it is more sinewy.

3, when the tendons of the muscles have been wounded, bad symptoms [result]. Convulsions, extreme pains [follow according to] Nicolò Massa;[216] therefore the tendons are avoided in [performing] a paracentesis.

3, the linea alba does not consolidate the sinews; therefore paracentesis in the fleshy part.

Veins and arteries from hypogastrium. Nerves more from the 6th pair, therefore extreme sensitivity; therefore dysuria or [pain] from a stone [or] old ulcer. Nothing [is] more intolerable than this, wherefore all desire death.

NB pain appears in the middle or end of the penis, therefore they desire to have the penis cut away. Therefore

[213] *Ibid.*, p. 798.
[214] A reference to the suprapubic opening of the bladder in cases of vesical calculus. The suprapubic operation was first described by Pierre Franco in 1556.
[215] *H.A.*, 519b:15: "The bladder, like ordinary membrane, if cut asunder will not grow together again, unless the section be just at the commencement of the urethra." This opinion is not true, for a bladder wound will heal readily provided there is adequate drainage of urine. A suprapubic opening into the bladder performed in Harvey's day would remain open by the continual drainage of urine. This would therefore be a basis for Harvey's acceptance of Aristotle's statement.
[216] Massa, ff. 12ᵛ-13ʳ: "You will also note that the didymi, which come from the substance of the peritoneum, perforate those chords of the muscles, whereby in the descent of the intestines or zirbus, often a laceration of the chord occurs: therefore let those who attempt to treat this open their eyes lest by violent intromission a laceration of the chord occur, since terrible results follow, so that there are pains, spasm and even death."

69

WH Δ compression on the perineum, pain ceases; cause may be in prostate.

The right spermatic vein [is] below the origin of the 23r emulgent, arises from the trunk of the vena cava in the superior and anterior part from a long, thickish root to which, [according to] Galen, there are sometimes branches from the emulgent.[217] To avoid the motion of the aorta, that on the left arises from the inferior part of the emulgent. Therefore male children from the right, female from the left, because more purified, perfect function [in the former].[218] WH X but not because the artery from the emulgent; not because the blood of the emulgent [is] less concocted; perhaps because the right side more perfect.

NB WH but more exactly about it afterward; it is sufficient to say now that they are here; here they run within the tunics and obliquely *on this Ridge;* here they cross the ureters.[219] Here with a nervule of the 6th pair and the cremaster [muscle] they descend into the testes,[220] going forth from the venter. Then returning by the same course,[221] the deferentia here enter the venter; here they descend into the inferior part of the bladder and terminate at the prostate.

Hitherto of the site, afterward of the remaining things.

[217] *De Usu Partium*, XIV, vii; Daremberg, I, 107.

[218] Harvey persists here in the ancient idea that a male foetus is produced by the sperm from the right testis and a female from the left. He does not actually mention the testis here but that is probably implied. This idea comes from Aristotle, for in his *De Generatione*, 763b:25, he says: "It is disputed, however, whether the embryo is male or female, as the case may be, even before the distinction is plain to our senses, and further whether it is thus differentiated within the mother or even earlier. It is said by some as by Anaxagoras and other of the physicists, that this antithesis exists from the beginning in the germs or seeds; for the germ, they say, comes from the male while the female only provides the place in which it is to be developed, and the male is from the right, the female from the left testis, and so also that the male embryo is in the right of the uterus, the female in the left." Aristotle believed the semen was concocted from the blood, and this explains Harvey's reference to the spermatic veins.

[219] Refers to the relationship of the spermatic vessels and ureter at the brim of the pelvis.

[220] The cremaster muscle does not join the spermatic cord until it is in the inguinal canal.

[221] The vasa deferentia from the testes enter the abdomen and terminate in the seminal vesicles at the back of the bladder above the prostate.

The quantity of the venter from the contained quantity of fat and viscera.

Wherefore according to Aristotle [there is] in them much flesh and compressed and obscure veins, small viscera and likewise a small venter. Indication from the mouth; and they are separated far below from the umbilicus. Shape; nature does not appear solicitous of the shape. In long animals they all appear very long, as in the eel, *stote, serpent.* Round for the round as Rhombus [in] *plays, soals, crabbs.* NB NB Here must be noted 2 divisions of the abdomen; dangers of wounds in the gastric and epigastric regions.[223]

Thus poynted out the situation &c. These parts serve for 23v coction, sanguification and chilification; for propagation of the species: males and females. The mouth and gullet receive for chilification; the stomach confects; the omentum, pancreas, vein, artery and viscera aid; the small intestine distributes; and the large receives the thick excrement.

The mesenterics receive for sanguification; the liver and spleen confect;[224] the veins distribute through the forces of the body; the gall bladder as the uppermost part, liver, kidneys, ureters and gibbous bladder [receive] the excrement.

For generation: the male testes and parastati[225] receive, prepare and confect; the deferentia carry down; the prostate preserves; the penis excretes. Females: as many things in their places.

[222] Harvey was apparently referring to the *Articulations*, section 49, in which it is stated with reference to fractured ribs: "the ordinary regimen is sufficient. For, unless they be seized with continual fever, a strict diet does more harm than good, by inducing inanition . . . for moderate fullness of the intestines has a tendency to replace the ribs, while evacuation leads to suspension of the ribs, and suspension induces pain." Crooke, p. 170, like Harvey citing section 3, says: "And Hippocrates conceiueth that the extuberation or distension of the stomack at the orifice is not backward but forward; whereas he saith, That the repletion of the stomack is a direction for broken ribbes."

[223] Wounds in the epigastric and gastric area may possibly involve the liver and spleen.

[224] Galen regarded the liver as the source of the veins and indeed of the blood. This Harvey also believed; cf. n. 291.

[225] The term was used by Hippocrates to mean the epididymis. Herophilus and Galen called this the varicose parastata to distinguish it from the glandulose parastata or prostate. Rufus of Ephesus also called the Fallopian tubes the parastatae varicosae. Harvey here uses the term for epididymis.

Necessity of all these hence: because to persist without nourishment [is like] a candle [without wax]. Because [on] the contrary [if] nourishment [is] unconcocted, therefore [death will result].

Place of coction and accomplishment like earth, for plants draw out concocted nourishment. Animals carry the earth within themselves, therefore the venter the most common part; and nothing [exists] without [it] from *invisible flye to the Elephant*. Where [there is] nourishment [there are] these four faculties: [1] attraction, [2] retention because time [is] necessary for coction, because nothing is permitted in motion. Likewise [3] digestion of what has been concocted wherefore there is present likewise [4] expulsion of excrement, because [it is] contrary and corruptive; where there is coction there [is] a remaining excrement. Therefore by these four all [have] health: ingest, extract, expel.

Since these four faculties with time and arrangement of 24r those things which ought to be attracted, concocted, retained and excreted, necessary that nature constitute different parts for these separately for their different functions. Because the same part in the same time could not possibly do all things or some, since they are contrary. Wherefore in all animals 1st, the mouth as a necessary place of ingress; 2nd, egress, viz., the anus [which is] opposed and different. Opposed because retention at the same time of what has been concocted so that it may nourish, of that not concocted that it may be concocted, and of excrement that it may be expressed freely, is inconvenient and impossible of consensus in the same place. Nature made a distinction which is better, and thus these three differ in a successive order so that they can not be confused.

Therefore nature [has made] different successive parts: stomach, liver, spleen for coction[226] or retention when [nourishment] is concocted. Small intestine and veins and arteries as a receptacle for what has been concocted and for

[226] *P.A.*, 670a:20: "The liver and spleen assist, moreover, in the concoction of the food; for both are of a hot character, owing to the blood which they contain."

distribution. The colon, rectum, bladder &c. for excrement.

What has been said refers to the first concoction and the second. But iij concoctions. For the more honorable and perfect animals, more perfect food; but more things required for the more perfect. So for coction *more refining more dilligent*. And *divers places* diverse heats, thus.

Wherefore by nature *divers offices and divers instruments;* 24v just as by chemistry in an elixir *divers Heates, vessells, furnaces to draw away the phlegme, rayse the spirit, extract oyle, fermentate and prepare, circulate and perfect; soe Nature* [has devised] the mouth of the stomach, mesenteric intestines, liver &c.

Account of all the operations of nature &c. &c., see p. 5. New: that contained preternaturally; dissection is useful for physicians; this more for surgeons for [relief of] the suffering. Therefore that which is in the dead body [should be studied].

In all dropsy [there is] water and flatus, wherefore diverse opinions. **WH** I saw: in a different way. Once dissected liver with mucous as Fernel [describes].[227] Some [believe] through the tunics of the intestines, that it returns through purgation. **WH** I believe that by diapedesis [one can get spread] from an empyema [of the lung to] the liver; through the mesenteric vein; through the epiploic vein; kidneys lax; [according to] Plater it is perforated;[228] for they purge through the intestines and a kind of vapor after densation. See the vesicles from the peritoneum *like frog spaune*. See the very wide vein in the hare as they say is received in the loins always.

Water of dropsy like lye as in all excoction *parboyled*. Stomach and intestines as in fox in shop. Mesentery without blood. Liver *grey, retracted hard splen watchet* with cartilage;[229] large blackish gall bladder; fat *curded*. Kidneys livid, very hard, lax without blood.

[227] *De Partium Morbis*, in *Universa Medicina* (Paris, 1567), p. 293.
[228] Plater, p. 154.
[229] Here Harvey is describing the post-mortem appearances in a case of abdominal ascites which, from his account of the liver, resulted from cirrhosis. A cirrhotic liver can well de described as "*grey, retracted, hard*."

Sometimes liver *Iron grey* large; large spleen; weakened intestines. Water in some: *sharper and milder*. Wherefore with such the outcome of paracentesis [is] death; wound *gangren* easily.

I saw such variety that I readily believed rather that cor- 25r ruption of the liver occurred from dropsy than the contrary, just as dropsy rather than scurvy at least in some. For from errors in diet [arises] cacochymia,[230] which persevering [produces] cachexia from which [one gets] corruption of the viscera.

See before timpanites, as a diabetic passion, hot, thirsty, somnolent;[231] foaming *saliva one month before;* wherefore I have cured diabetes, dropsy, and timpanites of this kind by refrigeration and mollifacients after other things [have been] to no purpose, as the Arabs [do] by [using] camel's milk.[232]

Soe yt this disease beginning in manner of timpanites or ascites, *they by much drinking fill their body which the kidnes not avoyding,* they augment themselves from the nourishment; finally, indeed, fleshy particles collect and are turned into water, and are carried away, as I have frequently observed; as long as some of the liver remains they live and do not die. Furthermore *all thes decay* at the same time proportionately during the dropsy; furthermore see the liver consumed without dropsy. WH In a dissection as far as the horns of the uterus I saw all the blood to have collected in the capacity of the abdomen through diabrosis.[233]

[Blank]   25v

The use of the intestines [is] to retain concocted aliment 26r

---

This is a very early, if not the earliest, description of cirrhosis of the liver. The credit for priority in description of this condition is usually given to John Browne (1642–1702), who described a case in 1685 in the *Philosophical Transactions* of the Royal Society. Harvey's brief account should now receive its due priority.

[230] A depraved state of the humors, also mentioned on f. 14ʳ.
[231] Typical symptoms of untreated diabetes.
[232] In the seventeenth and eighteenth centuries diabetes was treated on purely empiric lines and milk was a favored article of diet for this condition.
[233] This means ulceration or, more accurately, corrosion.

when it is extracted from the mesenterics [and] to complete coction in the excrement remaining. Aliment having entered the mouth begins at once to be changed [and] is constantly altered until it makes exit through the anus.

More [coction occurs] there where it is delayed longer; as in a confined place when the place is hotter, fleshier or by nearness to the hotter parts of the liver which promote coction; for pepper within, poultices without [promote coction]. Therefore not only the small [intestines], but the large [concoct even] more. The colon beyond others [because of] delay by the thickness of the matter and because of nearness to the liver, omentum, spleen; therefore an injected clyster [is] more nourishing.

Shape. Not a reservoir, but the concocted aliment [moves] in the intestines just as in little streams: 1 lest it putrefy; 2 lest it pass out through the exterior part of the vein. Not from the lowest because of frequent envelopment, and that which first was within prevented from protrusion exteriorly by the orifices of the veins.

Large and small [intestines are present] in certain animals, not in all: *weesel*, porpoise, all birds, one kind, but large because they concoct when some nutriment yet remains. Because they retain hard excrement and flatus; therefore in those in which there is no colon, no gurgling of the venter. [According to] Aristotle in no birds except the turtledove.[234] Nevertheless in the sting ray in which no gurgling there is a loose and fleshy intestine which is employed as a second stomach.

The duty of the large [intestine] more to detain excrement, of the small, chyle lest we be compelled to void continually. The duty of the colon for the greater amount of flatus and excrement in man. Shape of the colon therefore large, thick, fleshy, strong, with numerous cells,[235] *which a*

26v

---

[234] *H.A.*, 633b:5: " . . . but some birds have a peculiar habit of making a noise at their hinder quarters, as, for instance, the turtle dove; and they make a violent movement of their tails at the same time that they produce this peculiar sound."

[235] This is a reference to the sacculations along the length of the colon produced by the taeniae coli being relatively shorter than the bowel itself.

list gathered; elegantly in Du Laurens;[236] except where it is connected to the spleen and is lacking; *list in the top and fore part* wherefore small bits, grape skins, *plumstons* through the anus. *Likewise leaden Bulletts*[237] *gowld ring: mine.*

Huge colon in father;[238] Δ *Mr. Shaw*

The duty of the caecum to store up, to preserve, and concoct such concocted nourishment, and according to Piccolomini[239] lest any liquid nourishment be lost but be saved just as in the case of a juice which has flowed away. Wherefore in some a second stomach, as I have clearly determined, as in the mouse, the rabbit, *ginneycunney;* very ample and long for chyle, filled with longitudinal fibers; in the hare much greater and thicker than the [first] stomach, filled with very soft faeces. In the mole no vestige of a caecum, neither in the *stote* nor the seal.

Difference of caecum from colon because the former [is] monoculus as *the crapp;*[240] but the colon [has] ingress and egress.

Swollen aposteme equal to the stomach. Riolan Δ appendices to the ileon and the caecum;[241] [the latter] not of size of ileon; worms in caecum.

In the human foetus it also receives liquid faeces and it is like part of the colon X WH Δ. In man worthy of note as the papillae. The appendices in fowl as *goose duck lapwing &c;* *goose* around the end discolored by liquid excrement; why it may be absent in the ostrich I do not know; long corrugations are inserted near the anus almost to the rectum intestine as in sting ray, as also in *yellow Ammer.* They are not in *Hedg sparro smale birds woodcock;* thrush, blackbird; in fish but near the stomach; in some many as the *whiting*

---

[236] Du Laurens, p. 277.

[237] Cf. f. 20ʳ.

[238] This note must have been made after the death of Harvey's father on June 12, 1623, for the observation could only have been made at autopsy.

[239] *Anatomiae Prelectiones* (Rome, 1586), p. 86.

[240] Probably used here for crop, i.e., of birds. Cf. f. 30ᵛ.

[241] Riolan, p. 173: "Place the origin of the ileum where the intestine begins to be narrow and livid, for this intestine appears narrower than the rest; but it is more than twice the length of all the other intestines, and has very few mesenterics."

hering red fish whiting and mopp; some in a cross, marine frog, *like a tuffe of grass-rootes;*[242] cartilage; [according to] Aristotle none; X WH the sting ray worthy of note.

Valves: some affirm, some deny [their presence]. WH Δ it inflates [i.e., valves cause inflation] [and the bowel content] may be retained, as you see. A great quantity of water may pass through [the intestines] if injected. It may pass through Δ in the living, and sometimes in the dead;[243] in the body, outside the body more easily. Because *clister* delivers through the mouth, also placed below tied by a string to the legs.

Uses: lest [the content flow] back and because the colon [would] then [contract] upward and because of flatus.

WH These things rare so that usually [do] not [occur]: liquid flatus; neither in the living nor the dead. WH Those who say, as Sal[amon] Alb[erti][244] that there is within [the bowel] a membrane which closes the passage as in the veins X. Those who deny an impediment to it X. Because the interior tunic of the ileon is corrugated as if a valve opposed as at a *Hens-ars.*

Length; NB WH of length see before what [was said] on the site of the ileon. Capacity proportional to the stomachs

27r

[242] The presence of gut-appendages or caeca in birds and fishes is well described by Aristotle, on whom Harvey's passage is based. *H.A.*, 508b:15–20: "An exceptional property in fishes and in birds for the most part is being furnished with gut-appendages or caeca. Birds have them low-down and few in number. Fishes have them high up about the stomach, and sometimes numerous, as in the goby, the galeos [the dogfish; this has no caeca,] the perch, the scorpaena, the citharus, the red mullet, and the sparus; the cestreus or grey mullet has several of them on one side of the belly, and on the other side only one. . . . Some fishes are entirely without the part, as the majority of the selachians. As for the rest, some of them have a few and some a great many." *Ibid.*, 509a:20: "The gut appendages or caeca in birds, as has been observed, are few in number, and are not situated high up, as in fishes, but low down towards the extremity of the gut. Birds, then, have caeca—not all, but the greater part of them, such as the barn-door cock, the partridge, the duck, the night-raven, the ascalaphus [possibly an owl], the goose, the swan, the great bustard, and the owl. Some of the little birds also have these appendages; but the caeca in their case are exceedingly minute, as in the sparrow."

[243] Presumably through a defect of the ileo-caecal valve. This is rare, but F. Parkes Weber collected and published a number of cases of faecal vomiting (*Med. Press* [1950] 223:110–116). Cf. n. 140.

[244] "De valva ilei foribus ad introitum in colum praefecta," in *Tres Orationes* (Nuremberg, 1595), f. [95ʳ = M7ʳ].

and body and, on the other hand, the body to the intestines; wherefore *dwarfing of dogg with dasy rootes*[245] *as trees in potts: dwarf: cherry.*

Motion: as they detain so they distribute and drive downward. The inferior part opens itself by straight fibers; the superior by transverse contractions downward; as the *pudding wifes hand.*[246] Example of motion is snails and leeches. Undulant motion as worms, wherefore mica placed on the shell is moved; as *whip*; wave *in the water*; a cloth now fixed in all parts now movable. It appears *hawke putts over her meate crap turning vpon the giserd erecte itselfe.* Likewise *a dog: by the fier: cunny in ye sunn.* Rectum intestine which thus descends that it ought to be perceived by touch; often it is impeded by an injected clyster: wherefore it is opened by perspiration.

[Muscle] fibers [are] for the sake of motion, *sinewes,* 27v *threeds.* Straight: [1] long longitudinally opening by contraction [2] transverse compressing by contraction. The oblique are unable to detain by tonic motion, because [according to] Fallopius the straight are opened and the transverse are not compressed, by contraction.[247]

[245] Cf. J. Gerard, *Herball* (London, 1597), bk. 2, ch. 93: "The juice of the leaves and rootes . . . given to little dogs with milke, keepeth them from growing great."

[246] Cf. n. 247. This homely simile describes perfectly what happens. "Pudding" is a kind of sausage, as in the modern English "black pudding." As the *pudding wife* fills the skin, she squeezes the contents with her hand in order to press them down to the bottom of the skin; that is, transverse pressure is converted into longitudinal motion.

[247] *Fallopius,* ff. 170ʳ–71ᵛ: " . . . I speak now of the fleshy fibers which contract themselves—and that the transverse are for expulsion, since in my opinion all expel. . . . But if the more prudent were to reply, they would perhaps say that these fibers retain when they act and are contracted; you may say that when the oblique fibers are contracted, there is no doubt that they do so obliquely and that when the transverse are contracted, they act transversely. Then these do two things; first, they narrow the cavity transversely, and second, expel transversely; and accordingly the oblique narrow the space obliquely and expel obliquely. Therefore, willy-nilly, they do not retain. But if, on the other hand, they were to reply that since these fibers retain when they are not strongly but very moderately contracted, so that they scarcely grasp the food and thus retain it, you would say in contradiction that the transverse fibers do the same when they act moderately, and because they scarcely grasp the food, they retain in the latter way. This is also seen in the case of the hand when we squeeze mud or any other soft body with the fingers. If they are moderately contracted, they retain but if they act strongly, they express and expel. Hence we conclude that in

Straight fibers are more conspicuous in the rectum intestine and stomach, but the stomach, gullet internally and the intestines externally [display] few; in the slender [intestines] (because they detain more) [according to] Bauhin they strengthen the transverse so that in parts of the rectum the transverse [contract] more swiftly;[248] many in the colon, most in the rectum.

The transverse [are placed] outwardly in the stomach and gullet, but in the middle in the intestines. The oblique internal are distinguished by dissection [and are used for] retraction, especially coction. With the use of all diligence WH I have not found [in] the stomach *more coates than I made*, wherefore I believed the membrane of the vein not like woven straw, but like *silke-worms webb and spiders: snayle;* Fallopius therefore says all the fibers [are] in the same membrane;[249] thus [there is] a great controversy over the site and number of the fibers. Likewise of the membranes and tunics wherefore [according to Du] Laurens all membranes [are] double.[250] WH Multiple and infinitely divisible as I stretch. Yet *in tripes* they appear so evident that they resemble slender muscles. Wherefore WH in the first formation bones, flesh &c. similars; so the intestines afterward [are] in some divided and perfected. As appears

---

the stomach the function of retention does not rest with the oblique fibers but with the pylorus itself which constricts and obstructs the exit, and, in addition, the oblique as well as the transverse fibers join in this task, but—so to speak—in negative fashion, because they remain quiescent on the food and do not express it."

[248] Bauhin, p. 125: "They have fibers of all kinds; the interior, oblique that it may retain; the middle, transverse that it may expel; a few straight are added for protection of the transverse lest they be torn apart; so outwardly the straight are gathered together as in a ligament. . . . But there are very few straight in the slender [intestines], more in the thick."

[249] *Fallopius*, ff. 159ᵛ–60ʳ: "The whole internal tunic of the stomach is nerve-like, and vessels are led through and terminate in it. It possesses all types of fibers, straight, oblique and transverse, but, as in other nerve-like membranes, woven together and not visible without tearing apart so that it may better hold liquid substances and not leak like a sieve, and also that it can expand in all directions . . . a covering perfectly interwoven of straight, transverse and oblique threads."

[250] *Du Laurens*, p. 303: "The substance of the stomach is membranous, formed of two special tunics, and a third common one, almost innumerable branches of veins and arteries as well as many nerves."

in the muscles of the arm, so in the cranium and skin on top, wherefore nature imitating the artist, makes the delineation and projection and afterward completes and colors.

Three tunics from three kinds of fibers: 1 from the peritoneum X WH, 2 more special. External smooth, thin as I have distended from the peritoneum, but in the gullet fleshier as appears *in ye Butcher*, wherefore round muscle. The middle, fleshy from the parenchyma of the intestines. The internal very hard in the stomach because of the nourishment for the fleshy intestines; *like vnshorne velvet* with a covering; in brutes *Honycomb*. These equipped with all kinds of fibers; they are very powerful WH because [there are] clefts everywhere. 28r

When mucous of the intestines is scraped off; muddy fat appears [in the stools] in tenesmus and by a clyster in sharp fluxes. The interior tunic having been doubled in folds covers the mesenterics with cushionings and crusted mucous. *Hooping the gutts:* wherefore in these corroded [by] slight dysentery [or] by ulcerations of the intestines; by descent of humors from the mesenterics; it folds downward wherefore the transit of foods [is] downward; but upward *agaynst the heare;* wherefore defect of them, vomiting, *Clister;* but in the gullet the internal tunic without folds &c., wherefore here and there it passes through; in the horse [the same] as appears in ruminants.

Composition. Intestines therefore [made up] of tunics and these from fibers, flesh, parenchyma, veins, arteries, mesenterics, mucous: crust, fat.

Substance very similar to skin, partly fleshy, partly nervous, wherefore it is continuous through the interior like skin, wherefore similarity to skin in cold colic, wherefore taste of clyster in mouth.

Some thicker, fleshier, others thin, diaphanous, as *gutts in the shopp wett partchment.* The former concoct better for they concoct flesh, the latter poorly, wherefore the flatulent and infirm [have these].

Wherefore wounds in the former are to be cured because fleshy things consolidate better; wherefore [accord- 28v ing to] Galen a wound in the fundus of the stomach is consolidated and nourishment is better concocted in the fundus.[251] Because [the fundus is] fleshier as X Vesalius,[252] [but cf.] Colombo,[253] in those in whom there are much mucous and fat.

[Mucous and fat] more easily endure abuses: *drink: sur-* Errors *feits* wherefore more easily do they endure medications; thus during life unless they employ a cleanser [the intestine becomes filled with] slime, mucous and fat, as channels by mud, and appetite diminishes. On the other hand, there are the emaciated and thin, without mucous and watery [secretions] in place of mucilaginous; sensitive to all things, they are readily harmed by the acrid, salty [substances] because they are exposed.

Drink; wherefore they vomit more swiftly [rather] than they are able to become drunk, for wine disturbs them as it does in some in the jejunum and stomach; on the other hand, in others it reduces and prohibits ascension of vapor [as] *fatt* [in] *pott.*[254]

Purge; hence when they dispel humors or mucous by purgations [they are] joyful. On the other hand, mucous consumed for 3 days before generation, *sickish, windye.* Because nourishment [is] dry their mouth [is] without pituita.[255] Hence venter corresponds to face; [witness] the drawn face of the infirm.

State of health [can be judged] from the face; because they are sensitive to alteration of all food, and because much flatus collects because of amplitude. Likewise they generate [flatus] and are unable to expel because of weakness.

Observations: worms [exist] especially at the end of the 29r colon. *noething so comon as wormes fish mawes* flatus and *Colick.* WH Not all colic from flatus. Because [they get]

[251] *De Methodo Medendi,* bk. 6, ch. 4; Kühn, X, 419.
[252] *Fabrica,* p. 418.
[253] Colombo, p. 418: "The second tunic [of the stomach] which [Vesalius] said was slightly fleshier, but nevertheless is not more fleshy in its fundus than in its upper part."
[254] I.e., as a layer of fat on a stew prevents the vapor from rising.
[255] Phlegm.

81

distention from flatus in tumors of the venter without pain and [also get] pain without distention in a tumor of the venter from stinging and pungent material; and [those things which] arouse flatus.

Hot inflammation of the intestines [produces] veins filled and distended. In color purple and blackish. Thicker in the tunics yet smaller in the cavity. On the other hand cold, white and livid with flatus without blood. Humid, relaxed *wett partchment or lether*; example of flatulence *cattle going to grass*. In horse *loose belly fundament swabby; gutts creake and wallop*.

In some more cold than humid [with] tardy digestion. Because of the supply of obstructed mucous and lack of sensation. All those as horses *faynt and without spirit*. WH Wherefore in diseases the bowel may conduct liquids; it is harmful to a healthy condition for it weakens.

On the other hand dry and without flesh; without mucous; *transparent*, small [intestine] flatulent; wherefore near death they are inflated underneath and die unless as in some livid and black mortification.

Worse gripes from obstruction of small [intestine]. Woe is me;[256] in fluxes of the humors, or here descending, biting ulcerations are produced. Poisonous adust matter.[257] Cancerous, of bad quality. WH I have seen [this] just as aphthae.[258] WH *Ratt: poysoned:* by glass [fragments] *long scratches inflamed.* WH *Curtesan padoa meals eaten* in the stomach.[259]

[Blank]   29v

Action of the coelia[260] [is] chilification: [converting the 30r

[256] An old term for pain in the intestines, analogous to "Lord have mercy upon me," according to Hadrian Junius, *The Nomenclator* (London, 1585), p. 433, was used by "the homelier sort of Phisicians" for "iliac passion" or "a paine and wringing the small guts." Cf. n. 140.

[257] "Adust" was a term applied to a condition of the humors accompanied by heat and dryness of the body, thirst, black or burnt appearance of the blood (which is deficient in serum), and an atrabilius or melancholic complexion.

[258] Aphthae are spreading ulcers in the mouth or on the lips which are caused by fungi.

[259] A courtesan of Padua, recalling the period of Harvey's study there.

[260] This term was originally used to mean a cavity in any part of the body or in any viscus. More commonly applied to the intestines.

food] *into a pappy substance;* wherefore when blood is prepared from chyle it is separated in the liver. Four faculties *to attract,* concoct, *expell* &c. *Attract* occurs by hunger before nausea takes hold; on the other hand nausea rejects; *taking of a purge.* Coction a work of nature and of innate heat;[261] wherefore X animals concoct [in] the stomach. W H Rather man [in] the stomach. Because the liver, spleen, omentum, vena cava, kidneys also fomentations; as of kittens; wherefore cat of the old man; *David* girl;[262] wherefore *catt a witch* because it sought *to yeong people.*

Hence [according to] Aristotle the ancients understood that heat not only [as] hot water,[263] because of tenuosity of heat to penetrate interiorly. Retention *whereby it closely imbraceth &c; equally mashed and grownd with the drinke;* that when it does not do this it causes flatus, *Clydon*[264] *as horses;* or because of preternatural food, wherefore vomiting without drink; and vomit of bile and pituita as in boys without food; wherefore all stomachic constricting things.

What coction is; according to the confusion of authors [a subject] filled with difficulties. Some [consider it] liquefaction; some, grinding and cutting and special attraction

Coction,
chilification,
first concoction
of the stomach;
quantity, shape

[261] *P.A.*, 650a:5: "Now since everything that grows must take nourishment, and nutriment in all cases consists of fluid and solid substances, and since it is by the force of heat that these are concocted and changed, it follows that all living things, animals and plants alike, must on this account, if on no other, have a natural source of heat." *Ibid.*, 650a:10: "After the mouth came the upper and the lower abdominal cavities [stomach and caecum], and here it is that concoction is effected by the aid of natural heat."

[262] I Kings 1:2, referring to the declining strength of David: "Wherefore his servants said unto him, Let there be sought . . . a young virgin. . . . let her lie in thy bosom, that my lord the king may get heat."

[263] *P.A.*, 648b:35: "In some of the bodies which are called hot the heat is derived from without, while in others it belongs to the bodies themselves; and it makes a most important difference whether the heat has the former or the latter origin. For to call that one of two bodies the hotter, which is possessed by heat, we may almost say, accidentally and not of its own essence, is very much the same thing as if, finding that some man in a fever was a musician, one were to say that musicians are hotter than healthy men. Of that which is hot per se and that which is hot per accidens, the former is the slower to cool, while not rarely the latter is the hotter to touch. The former again is the more burning of the two—flame, for instance, as compared with boiling water—while the latter, as the boiling water, which is hot per accidens, is the more heating to the touch."

[264] A fluctuation and flatulency in the stomach and intestines (Motherby, *op. cit.*).

by individual parts; others a kind of putrefaction and fermentation; others a distillation through descent and retort. WH All have spoken partly correctly, partly incorrectly, because [it is] not something wholly of itself alone. Therefore philosophical (for [it is] a philosophical dispute). Coction [is] change of the whole substance with generation and corruption. Chyle and blood from food and drink mixed. Chyle is first.

While this is happening it is ground and broken up. Be- 30v cause it is prepared by the teeth and the mouth and then distilled, as it were, by a hypothetical retort in which it is liquefied by the heat of the liver.[265] It is broken down in the same way as all generated colica is broken down; not before ought this be said to putrefy; because it has acquired a better foramen; name from the end; so that all things [are] in motion and about to move.

It is ground by the teeth which in some [animals are] in the fauces. WH As *Barbel roch* in some in the stomach itself, *lopster*.[266] and the teeth may be outward.

Coction of the stomach [is] the first concoction; [a stomach] is lacking in plants for they drink up concocted aliment from the earth. Nature very solicitous for this coction in animals, wherefore it placed [it] around the liver, omentum, spleen, &c.; teeth [used only] for preparation.

[Of] those which use crude aliment some [have] a large stomach. 1, wherefore [are] equipped with teeth in another part; *Chamel* &c., reticulated, omasum, abomasum;[267] wherefore their flesh more delicate than that of horses;

[265] *P.A.*, 670a:20: "The liver and spleen assist, moreover, in the concoction of the food; for both are of a hot character, owing to the blood which they contain. . . . The heart then and the liver are essential constituents of every animal; the liver that it may effect concoction, the heart that it may lodge the central source of heat."

[266] *P.A.*, 679a:35: "The Crustacea also, both the Caraboid forms and the crabs, are provided with teeth, namely their two anterior teeth. . . . Directly after their mouth comes a gullet, which, if we compare relative sizes, is small in proportion to the body: and then a stomach, which in the Corabi and some of the crabs is furnished with a second set of teeth, the anterior teeth being insufficient for adequate mastication." *Ibid.*, 679b: 1 n.: all Decapoda have stomach teeth.

[267] The four stomachs of a ruminant animal are first the ventriculus, second the reticulum, third the omasum, and fourth the abomasum.

84

wherefore in the sacred scriptures the ruminants not *vncleane*. 2, wherefore birds [have] a crop, *crap*, and *gisard*;[268] furthermore, a fleshy stomach compensating for teeth; nevertheless *little Birds* in which a cartilaginous stomach of slight heat and good force. *Goose duck cormorant* &c., with wide gullet,[269] wherefore *ther flesh not delicate*; 3, of those with teeth in both parts, some employ the caecum or the colon for a second stomach, or they do not require aliment so carefully sought out as horses, or as carnivores desire prepared aliment.

Grampos, porpos have four stomachs in regard to order, 31r size and structure. The elephant is said to have a small stomach because it is compensated by the length of the intestines. In some the upper part of the stomach is constituted differently, wherefore in the single [stomach] a more fleshy fundus [compensates for] a second stomach. Man as in other things is supplied with intellect and skill; with skill in cookery, for he employs cooked foods.

Size of the stomach. Stomachs of certain gluttons, *Gourmandisers, drinkers* have been of huge capacity; many ancient and recent accounts [of these]. From Vopiscus [in regard to] the gourmand of Aurelian *favorit eate a Boar a mutton and a pigg.*[270] WH *Wilkinson of Cambridg. pigg of ye spitt.*[271] Milo of Crotona 20 lb. meat *att a meale eate the bul-*

[268] In the manuscript the word *ingluvies* is used, which is the Latin for crop. After this, Harvey writes *crap* and *gisard* as the English equivalents. His use of *crap* instead of crop will be noted. Strictly speaking, the terms crop and gizzard are not synonymous, the crop being an outgrowth from the gullet, and the gizzard being the lower, pyloric portion of the stomach.

[269] *H.A.*, 509a:4: "The duck, the goose, the gull, the catarrhactes, and the great bustard have the oesophagus wide and roomy."

[270] Flavius Vopiscus of Syracuse, "The Deified Aurelian," in *Scriptores Historiae Augustae*, tr. David Magie, III (London, 1932), xxvi, 54 (p. 293): "[Aurelian] had the greatest delight in a gourmand, who could eat vast amounts to such an extent that in one single day he devoured, in front of Aurelian's own table, an entire wild boar, one hundred loaves of bread, a sheep and a pig and, putting a funnel to his mouth, drank more than a caskful."

[271] Ralph Wilkinson (*ca.* 1544–1609), M.D. (Cantab.), Fellow of Trinity College. He was a Fellow of the College of Physicians, and Registrar from 1605–1608. He preceded Harvey in the office of Physician to St. Bartholomew's Hospital. Harvey, consequently, would be aware of Wilkinson's prowess at the table.

the next day.[272] *Athlete:* [according to] Aristotle regarding the supply of nourishment at the table;[273] are said [to require] 20 lb. *Att Hollingbore he yt eats Bollough livers. Of Drunkerd* no less fame. *Alexander sett a prise, many fell many died.* Promachus the victor; is said to have drunk four congij,[274] that is *20 quarts and better.* Torquatus tricongius;[275] Bicongius, son of Cicero;[276] Syracusius *drinke vntill an egge hatched;*[277] Persians: *games matches and prises for drink.*[278] Romans: how many contests *in ther Mrstris name;* many others; the account of both.

Scarcely a wholesome habit *to stretch* the stomach. As 31V WH I have seen the ureter [stretched], and Indians [to stretch] an opening in the ears; but the *ordinary* size of the stomach [was regarded by] the Greeks [to be best], 1, for

The stomach appears larger in those in whom the distance from the umbilicus [to the ensiform cartilage is greater] than that of the ensiform cartilage to the root of the neck. [According to] Riolan stomach of Ethiopian not larger than intestine[279]

[272] Athenaeus, *The Deipnosophists,* X, 412; C. B. Gulick trans., IV (London, 1930), pp. 369–371: "Milo of Croton, as Theodorus of Hierapolis says in his work *On Athletic Contests,* used to eat twenty pounds of meat and as many of bread, and he drank three pitchers of wine. And at Olympia he put a four-year old bull on his shoulders, and carried it round the stadium; after which he cut it up and ate it all alone in a single day."

[273] *G.A.,* 768b:30: "For owing to the quantity of their food their nature is not able to master it all, so as to increase and arrange their form symmetrically."

[274] Athenaeus, *op. cit.,* X, 436; C. B. Gulick trans., IV, 479, "[Alexander] . . . instituted a contest in the drinking of unmixed wine. . . . Of those who drank the wine, thirty-five died immediately of a chill, and six others shortly after in their tents. The man who drank the most and came off victor drank twelve quarts and received the talent, but he lived only four days more; he was called champion [i.e., Promachus]."

[275] I.e., Torquatus with a capacity of three *congii*—according to Harvey, 15 quarts.

[276] The orator's son gained the reputation of being the hardest-headed toper of his day, but according to Harvey he appears to have been only a two-*congii* man.

[277] Presumably a reference to Dionysius the Younger, Tyrant of Sicily, who was said sometimes to have been drunk "continuously for ninety days" (Athenaeus, *op. cit.,* X, 435; Gulick trans., IV, 473).

[278] Mithridates the Great, King of Pontus, once set up a contest in eating and drinking; Mithridates took both prizes (Plutarch, *Symposiacs,* I, vi, 2); according to Appian's testimony, drinking contests were the custom of Pontus (*Herodotus,* I, 172); according to Xenophon, the Persians drank to such excess that after meals they had to be carried out (*Cyropaedia,* VIII, viii, 10). The boast of one Persian king is reported by Athenaeus, *op. cit.,* X, 434; Gulick trans., IV, 469: "I could drink much wine and yet carry it well."

[279] Riolan, p. 197: "Before many spectators I demonstrated the stomach of an Ethiopian woman which from the oesophagus was not larger and wider than the intestine, although it was whiter and thicker."

necessity, 2, for health, 3, *pleasure*, 4, undisturbed sleep for father; in addition, for drunkenness, fury, insanity &c. In those in whom the stomach is large a sign [that it is extended] farther than the ensiform [cartilage] and the umbilicus, and on the other hand, in those voracious [the stomach springs into the mouth].[280]

Small size of stomach: [witness] German girl [who] fasted for iij years.[281] [According to] Galen smallness [of mouth] of stomach a disease[282] perhaps [caused] by restricted and moderate diet; *like a Horne a Bagpipe*. Round for greater capacity; *less and less* where cooked food [is eaten, for it needs] a smaller space.

Two orifices (cardiac mouth; pylorus four fingers from the fundus) and [twisted] upward, lest it be opened by weight of food. Pylorus more compressed NB *a shilling stuck which he always carried in his purs*.[284] [According to]

Shape

[Stomach of] ostrich partly, as [in] quadrupeds, membranous; partly, as [in] fowl, fleshy

[According to] Riolan twin stomach[283]

[280] Galen, *De Facultatibus Naturalibus*, III, 8; Kühn, II, 174; Brock trans. (London, 1928), p. 271: "In those animals, therefore, which are naturally voracious . . . when they are sufficiently hungry and are pursuing one of the smaller animals, and just on the point of catching it, the stomach under the impulse of desire springs into the mouth. And this cannot possibly take place in any other way than by the stomach drawing the food to itself by means of the gullet, as though by a hand."

[281] This case was widely noted in medical literature. Her name was Margaret———, and she was born in 1536 at the village of Roed, near Spire (Germany). The length of her fast varies in the different accounts. The first of these was G. Bucoldiani, *Brevis enarratio de puella quae sine cibo et potu per aliquot annos in pago Roed egit* (1543, reprinted 1587), which was followed by a tract by Simon Portius, *De puella germanica, quae fere biennium vixerat sine cibo, potuque* (1551). The most likely source of Harvey's knowledge of it seems to be Johann Lange's *Medicinalium epistolarum miscellanea* (1554). There were several editions of this work where no. 27 is entitled "Inedia triennalis puellae Spirensis."

[282] Galen, *De Locis Affectis*, V, 7; Kühn, VIII, 345–347; Daremberg, II, 646–647.

[283] Riolan, p. 194: " . . . in a town two leagues from Paris a woman was dissected in 1624 before thirty spectators. Her stomach was longish, compressed in the middle, with an amplitude equal to that of the colon intestine. Upon dissection I discovered that narrow part, like a pylorus, to end in another cavity which afterward was terminated in a very thick orifice, which opening was the pylorus, from which the first intestine began."

[284] Galen, *De Facultatibus Naturalibus*, III, 4; Kühn, II, 154; Brock trans. (London, 1928), p. 239: "[The retention of food in the stomach is not due to] the lower outlet of the stomach being fairly narrow. . . . One person . . . . swallowed a coin . . . yet . . . easily passsed by the bowel what [he] had swallowed."

Plater here fat a cause of lack of appetite.²⁸⁵ Very fleshy [according to] Galen,²⁸⁶ X glandular [according to] Galen X but in dog. Both orifices fleshier and thicker to touch and sight than the stomach. WH NB hence those things must be inserted [into the stomach] which first in situ.

[Blank]   32r

Viscera bipartite because the body [is divided] into 32v right and left. Some [viscera] clearly [separated into] two, as kidneys. Some not two but one bipartite, as the cerebrum, lung.²⁸⁷ Some imperfectly, the heart.

Liver and kidneys two viscera partly divided, wholly divided as kidneys. Because the same arrangement is seen [in both], because the same, and because partly a different arrangement; therefore great controversy and doubt regarding their actions. Wherefore WH as the arrangement partly the same [and] partly different, so the action partly the same [and] partly different; and the arrangement is more like than unlike, for thus

1, same arrangement viz., in vessels and principal parts; according to Galen parenchyma²⁸⁸ not different in a perfect liver than an imperfect one in another animal; wherefore false liver. Furthermore, they correspond in site. The liver in the upper part, right, forward, more noble place for the more noble [organ]. *Splen* on the other hand, inferior, to the left, to the rear, ignoble place for the more ignoble [organ]. And thus as much for place as for structure, size, color; wherefore false liver consequently differs as right from left.

2, nevertheless there is a certain difference in all in color,

²⁸⁵ Plater, p. 150.
²⁸⁶ *De Usu Partium*, IV, vii; Kühn, III, 280–281; Daremberg, I, 290.
²⁸⁷ Harvey still carries on the medieval conception of a single lung divided into two portions by the mediastinum. The brain has two cerebral hemispheres, but his "bipartite" argument has more force here than when applied to the lungs.
²⁸⁸ Galen, *De Facultatibus Naturalibus*, I, 6; Kühn, II, 13; Brock trans., (London, 1928), p. 23.

substance, vessels, size. Spleen darker, smaller, rarer, laxer, more abundant in arteries; wherefore spleen seems hotter than liver.

Function of the liver: a second coction, sanguification. 33r Question between physicians and philosophers and among physicians [regarding the formation of natural spirits in the liver]. Bauhin's opinions regarding the heart, liver, veins; [also] regarding veins and liver. [Du] Laurens: labor of the veins, color of liver;[289] and X in fish in which the liver [is] yellowish.[290] The more judicious physicians do not differ from the philosophers in the matter.

1, [according to] Aristotle the heart is the origin of all things,[291] therefore of sanguification; nevertheless he says

[289] Du Laurens, p. 239: "The stomach like a servant provides food for this liver; the gall bladder, spleen, kidneys purge the filth of this royal house as if they thrust it out of the kitchen. . . . It is the center of the dominion and the site of the vital and animal faculties." Ibid., p. 240: "Of the veins of the liver, some transport the thinner portion of the chyle from the hollow of the liver . . . ; they concoct, attenuate and prepare this portion; others lead the blood now confected and elaborated to the trunk of the cava. These roots of the caval and portal veins are sprinkled through the whole body of the liver and are so remarkably involved that most of the roots of the portal enter through the hollow of the liver, very few through its convexity; so that it is very likely that sanguification occurs in the hollow and distribution and perfection in the convexity. The anastomoses of these roots, unknown to the ancients, must be admired, through which all the veins in the liver, just as in its matrix, have communion, so that it ought to be called the origin of all the veins. Furthermore, nature constructed this plexus of the veins in the liver for the perfection and elaboration of the blood, so that because of the long delay in those compressed spaces of the vessels it might be more perfectly concocted and be altered through even the very smallest contact of the parenchyma on all sides. and so the coverings of the veins which are scattered through the flesh of the liver are very thin." Ibid., p. 239: " . . . the color of the whole surface is taken directly from the constitution of the liver; for such color as blossoms in the skin is taken from the humor; but the liver is the prime workshop of all the humors."

[290] In fishes the liver has a yellowish tint or is definitely yellow. Aristotle, P.A., 673b:30: "On the other hand, the liver of oviparous quadrupeds and fishes inclines, as a rule, to a yellow hue, and there are even some of them in which it is entirely of this bad colour."

[291] P.A., 670a:20: "The heart and the liver are essential constituents of every animal; the liver that it may effect concoction, the heart that it may lodge the central source of heat. For some part or other there must be which, like a hearth, shall hold the kindling fire; and this part must be well protected, seeing that it is, as it were, the citadel of the body." It is of interest that Aristotle states that the liver is not the primary source of the blood. P.A., 666a:25: " . . . no sanguineous animal is without a heart. For the primary source of the blood must of necessity be present in them all.

the liver [is] so necessary that [there is] no blood without a liver; thus [according to] Averroes imperfect in the liver, perfected in the heart.

2, physicians consider the liver, however, as the more influential faculty; wherefore the heart radically and fundamentally, the liver instrumentally; thus the eye sees [with the help of] the cerebrum, wherefore the cerebrum having been disturbed, [the eyes] do not see and they feel neither pain nor do they move.

Action of the liver: concoction, sanguification, but as the instrument of the heart and secondary; also useful for the first coction in the stomach, namely warming of the intestine and therefore blood filled with native heat. But native heat [is] the author of all things which concern coction and the instrument of instruments which first is made in the heart. [Liver] useful also as anchor by which the veins are made firm at its body. Thus the placenta in utero; it grasped the bitter liver.

To create
natural
spirits

WH *Yf I could shew what I hav seen yt weare att an end between physicians and philosophers* for blood rather the author of the viscera than they of it. Because blood present before the viscera, not coming from the mother, but from a spot in the egg. The spirit is in the blood. Native heat the author and where it prevails more, there [is] the first of the first. I have seen, as *Dr. Argent*[292] will be my witness, all things perfected with the liver unformed. The heart [is] formed with the auricles [when] the liver [is still] a rude and confused mass. The heart very white, the auricles reddish and filled with blood.[293]

It is true that sanguineous animals not only have a heart but also invariably have a liver. But no one could ever deem the liver to be the primary organ either of the whole body or of the blood."

[292] John Argent (*ca.* 1560–1643), M.D. (Cantab.), 1595, Fellow of the College of Physicians, 1597, and President from 1625 to 1633. To him, in his official capacity as President and as "the writer's particular friend." Harvey dedicated his *De Motu Cordis* in 1628, stating that he had "for more than nine years confirmed it in your presence by numerous ocular demonstrations."

[293] Here Harvey tells us for the first time of his observations on the development of the hen's egg, which was first published in his *De Generatione*

Similarly [according to] other physicians the *splen* [is] 33v
the receptacle of melancholy as the gall bladder of gall,
wherefore the *splen* causes to laugh: wanton *splene* WH by
which its constitution is more similar to the liver than its
function to the gall bladder. Wherefore recent writers
Rondelet,[294] Plater,[295] [Du] Laurens,[296] Bauhin,[297] Ulmus,[298] Vesalius.[299] It draws, retains and concocts melan-

(1651). His dogmatic statement on the primary development of the heart
is given here and, as he says, he demonstrated this to Dr. John Argent.
Aristotle, of course, knew of the early development of the heart (*P.A.*,
666a:10): "For the heart is the first of all the parts to be formed; and no
sooner is it formed than it contains blood." He also stated that the heart
and liver were to be seen in eggs as early as the third day (*P.A.*, 665a:35),
"being then no bigger than a point." Harvey, however, quite clearly states
that the heart is formed before the liver, and in this he was quite correct.

[294] For this and some of the following accounts of the spleen it is likely
that Harvey resorted to the convenient digests of them in Du Laurens,
p. 248, where we find it was the opinion of Rondelet that the spleen is not the
receptacle of the melancholy humor because as long as it is in its natural
form it is used for the generation and conservation of the bones and other
hard parts of the body; and because it is so small in quantity there is no
part designed specially for its reception any more than for the excrement
of the blood, which, for the most part, is consumed in sweat and insensible
transpiration. Rondelet gives his opinion directly in *Methodus curandorum*
(1609), pp. 515-516.

[295] Plater's opinion bears close resemblance to that of Du Laurens, recorded in n. 296.

[296] Du Laurens, p. 247: "Just as farmers place lupines around rich corn-
fields so that the corn may come forth more readily and sweetly from the
collected bitterness of the earth, so nature constructed the spleen in contrast
to the liver so that having expurgated the liver of its feculent filth and drunk
up the thick and dirty refuse of the juices, the blood may be rendered purer
and brighter. For this reason the spleen is called the organ of laughter by
Viaticus. . . . If the spleen is unable to perform its function of purifying
the blood, it is remarkable what grave symptoms appear and are sustained."

[297] Bauhin, p. 278: "Wherefore the spleen assists the liver in the production of blood, partly in so far as it produces blood of its own kind, partly
in so far as it repels the very thick juice unsuitable for the generation of a
very pure blood, so that better sanguification may occur in the liver; in this
manner it is correctly said to purge the blood, and to render it brighter;
hence the spleen was called the seat of laughter by the ancients, and it was
commonly said that the spleen causes laughter, the liver causes acrimony."

[298] The published transcription erroneously gives "Vlies" rather than
Ulmus, who wrote a small treatise on the spleen, *De Liene* (Paris, 1578).
Harvey probably learned from Du Laurens, p. 248, that Ulmus, or Umeau,
considered that the vital spirit was prepared in the spleen, that is, very thin
blood and matter of the vital spirit which was carried from there through the
arteries of the spleen to the left ventricle of the heart. There it was mixed
with air and perfected and distributed through all the arteries.

[299] *Fabrica*, pp. 439-440: "The use which seems most probable to other
professors of dissection teaches that it was constructed by nature as a

receptacle for the dregs and feculent part of the blood made in the liver, and just as the gall bladder is provided for the thinner and lighter excrement, so the spleen was constructed to receive the thicker and heavier; and that what the spleen, like a servant, draws and sucks forth to itself through the trunk of the portal vein it confects and elaborates and renders suitable for its own nourishment, making the blood, if it be thick and feculent, rare and spongy. They say that numerous arteries inserted into the spleen especially aid in this, greatly assisting the full elaboration of that blood by their heat. But because the primary and particular performer of this action may be considered to be the flesh of the spleen, even if I do not say it, I do not doubt that it is apparent. And so it has also been conceded by these authors that the spleen does not confect all the blood which it admits from the liver and that it stores it up as nourishment, and what is unable to be adapted to its substance it regurgitates into the stomach for some great use. For first all affirm that the melancholy juice is eructated from the spleen into the stomach, some, indeed, through the vein extending from the spleen into the stomach, others that through a special passage and then from the stomach into the intestines whence it is expurgated from the body with the faeces. Some write simply that that vein or passage is led from the spleen into the stomach, others boldly asserting the place of insertion say that it is implanted into the upper orifice of the stomach. A like dissension arises among the leaders of dissection in assigning the use of that excrement, for some assert that it is very useful and helpful for the functions of the stomach, just as yellow bile is unfriendly and very harmful; others are content to teach that this black bile by its quality of taste, which they confess to be sharp and acid, by its astringency and gathering force strengthens all the functions of the stomach which consist in embracing, and therefore prevents unconcocted food from slipping forth from the stomach. However, some not content with just this use, add to those things already mentioned that the appetitive force of the stomach is so incited by that excrement that, as they assert, the vein or passage inserted from the spleen into the upper orifice of the stomach—taught by imagination not by dissection—is for that special use.

"I dare affirm nothing regarding that eructation of the excrement of the spleen into the stomach nor, indeed, does dissection clearly reveal to me that which professors of anatomy very boldly assert without any argument. For first I do not believe that there is a passage similar to that by which the bile is purged into the intestines; and the veins which are implanted into the left side of the stomach go forth not from the body of the spleen but from those which are already inserted into the spleen. Nor is another internal course of those veins to be seen through the stomach differing from those seeking the body of the stomach; then there is certainly no vein here other than the unpaired artery. Finally, what is carried from the spleen to the stomach seeks the fundus of the stomach rather than its upper orifice, for the vein creeping through the left side of the fundus of the stomach is easily the largest of all the veins extended from the stomach to the spleen. Neither this vein nor any other approaches the left side of the stomach filled with blood different from that of the rest of the veins entwining the stomach, so that unless I were to fear the calumnies of many, and I have hitherto experienced very severe ones to my no little injury, arguments would not be lacking by which I might upset the common opinion regarding the duty of the spleen. Meanwhile, everyone has granted that the spleen by its innate heat and by the numerous arteries which are woven into it fosters the concoction of the stomach. Because it is commonly held that the spleen is the author of laughter, I conjecture that this has been assumed because it is believed to draw to itself the thick, feculent and dirty blood and, so to speak, render the rest of the blood more active and cheerful. I do

choly juice from the portals of the liver by the heat, laxity and number of its arteries. For the coction is not the melancholy but now the cold and dry part of the chyle.

2, According to Aristotle it is able to divert and draw the idle juices, rather than the sanguineous, from the stomach and concoct them.[300] Moreover, when the spleen has excessive excrement and little heat it is a small body weakened by too much nourishment. And because of the flowing together here of the humor, this diverticulum of the crude humor produces very hard venters.

3, Physicians differ regarding the melancholy juice, preserving the supposition of Galen taken from Hippocrates. [According to] Aristotle [it is] a crude, idle vapor.[301] Both consider that [it issues] from the stomach and liver because thick, earthy, cold and dry; and therefore not capable of concoction.

It is attracted and concocted by the spleen, wherefore the conclusion from these things which have been said expertly; and from these [things] which appear here in regard to structure and site. Both these viscera serve for sanguification. Because animals [are] like man, [with] foods for different parts, of which some are concocted easily, others with more difficulty. Some of good juice, others of melancholic, cold, dry. For [the food goes to] the liver and before it has overflowed at the portals (where also the bile is separated) it is diverted and attracted; there it can be received, concocted and perfected by the supply of heat and the rarity of the part.

Here the structure of the veins is demonstrated, for if it 34r receives through the splenic branch and from the portal of

---

not assert that by a special use of the spleen—as Aristotle considered the brain—that the heart is cooled, as the principal Arabs declare, since it does not touch the heart, and if any cold arises in it, the arteries of the spleen do not direct it to the heart."

[300] *P.A.*, 670b:5: "For the spleen attracts the residual humours from the stomach, and owing to its bloodlike character is enabled to assist in this connection. Should, however, this residual fluid be too abundant, or the heat of the spleen be too scanty, the body becomes sickly from over-repletion with nutriment. Often, too, when the spleen is affected by disease, the belly becomes hard owing to the reflux into it of the fluid."

[301] *P.A.*, 670a:30 ff.

the liver, so it is situated that it might receive and serve as a dependent of the liver. Wherefore they are very fortunate in whom the stomach and spleen are strong because they meet the first aliment just as among soldiers [there is a] *vant-guard*. Thus they are stronger, healthier; on the other hand, those in whom the contrary occurs are subject to difficulty and uncertainties. Wherefore splenetics [are] of two kinds.

1, with abundant crude aliment; they are deficient in heat for the spleen; large scirrhous spleens, very hard venters. Scorbutics.

2, on the other hand, with increased heat there is deficient aliment. [They are] melancholics, fanatics, flatulents.

Wherefore melancholics, hypochondriacs lack pleasant disposition, talent; much internal rumbling in them; ptyalismus;[302] in them *splen-stone* [seems] to me cold in its action, in those using a little thicker diet it is not abundant.

Wherefore when there is not that diversity of foods or it is accomplished in another way, there is only a liver; the spleen worthy of note, and according to the abundance or lack of this thick material the spleen is larger or smaller. The liver likewise by defect of the spleen is larger, and the opposite; wherefore when the liver clearly divided on each side, full hypochondria; there the spleen worthy of note; in the rabbit as if a lobe of the liver;[303] in the hare an ornament; it is lacking in all fowl because a fleshy stomach.[304]

2, divided liver large.

---

[302] A condition in which there is a frequent and copious discharge of saliva.

[303] *P.A.*, 669b:25–30: " . . . in animals that necessarily have a spleen, this organ is such that it might be taken for a kind of bastard liver; while in those in which a spleen is not an actual necessity, but is merely present, as it were, by way of token, in an extremely minute form, the liver plainly consists of two parts; of which the larger tends to lie on the right side and the smaller on the left. . . . Examples of such division are furnished by the hares of certain regions, which have the appearance of having two livers. . . . "

[304] *P.A.*, 670b:10: "But in those animals that have but little superflous fluid to excrete, such as birds and fishes, the spleen is never large, and in some exists in no more than by way of token."

3, [those that have] spongy lungs[305] do not take drink, for a copious drink of water makes crude aliment.

4, in fowl such crudity is consumed.

5, they use simple aliment; for seeds or flesh without drink.

See: *sparrowhawk*, kite, *owle*, pigeons,[306] *Goose duck woodpecker*.

[The spleen is present] in almost all fish and [is] notably 34v florid in *plays*, sting ray, *flowndors* and *Barble*, because in them such perfection of food is not necessary. The liver [is] a little more pallid, less serviceable and in place of the spleen, wherefore temperament vitiated; it is not [so] in seal and cetacea, for [they have] a placenta[-like structure] in place of the kidney WH.

In ovipara and quadrupeds [the spleen is] rigid and narrow, similar to the kidney because the lung [is] fungous; desire little drink and [secrete] into the outer covering as birds [do] into feathers. WH *land tortoys* liver spongy and blackish; spleen of vivid color as heart, kidneys. In *toad* it is of more florid color than the liver.[307] The liver almost semiputrid with a vitiated spleen; similar because temperament vitiated; they do not use choice aliment; therefore in place of liver a quasi spleen or a false liver.

WH Nevertheless in some quadrupeds [the spleen is] more florid and of better consistency which astonishes me, but I believe in those it does not serve a general use; but nevertheless worthy of note that in the larger the stomach,

---

[305] *P.A.*, 671a:10: " . . . all animals whose lung contains blood are provided with a bladder. Those animals, on the other hand, that are without a lung of this character, and that either drink but sparingly owing to their lung being of a spongy texture, or never imbibe fluid at all for drinking's sake but only as nutriment, insects for instance and fishes. . . . are invariably without a bladder."

[306] *P.A.*, 670a:30: "Therefore it is that in some animals the spleen is but scantily developed as regards size. This, for instance, is the case in such feathered animals as have a hot stomach. Such as the pigeon, the hawk and the kite." In his *Historia Animalium* Aristotle adds the falcon and the owl to this list. The spleen is small in all birds.

[307] *H.A.*, 506a:20: "With oviparous quadrupeds the case is much the same as with the viviparous; that is to say, they also have the spleen exceedingly minute, as the tortoise, the freshwater tortoise, the toad, the lizard, the crocodile, and the frog."

95

like the omentum, promotes its coction. *Catt, poulcat weesel.* Wherefore in men and in other animals in which the liver is larger, so a [large] spleen; on the other hand, in which the larger the spleen the smaller the liver and poorer in color; wherefore poorer condition; wherefore Trajan's remark regarding a tax.[308] Where the spleen more florid, and the liver similar to spleen in consistency and color, there it performs function of a servant.

WH I saw: a frightened person hanged on a ladder; *Cambridg;* spleen more divided into lobes, with small liver. Fernel describes a larger liver,[309] and Vesalius [describes one enlarged] from jaundice.[310]

[Size varies] generally [throughout the body] as [for example] the viscera, venter and veins; large viscera, increased venter, veins clearly filled with black blood; in them little flesh and on the contrary the flesh or fat of a florid color &c. The more it surpasses the viscera the more it lacks flesh and the opposite. Wherefore it appears that the liver and heart [are] necessary because of the origin of heat, for it is necessary that the lares and hearth be present, and for the sake of concocting food, wherefore there should be no lack of blood for those two. **WH** Nevertheless I saw [a man] live for months with a putrid liver; but I believe Benivieni saw a man without a liver but with an enlarged intestine,[311] as **WH** always *in a Barble* and partly *in a goose.* Here the liver a public region, part giving, receiving.

On the other hand the *splen* necessary per accidens be-

[308] Du Laurens, p. 321: "The spleen is not of one size in all as not of one color; however, in all the spleen is inconsequential in respect to its small size; and in those whose body is healthy the spleen is reduced in size; in those on the other hand in whom the spleen is augmented, the body is threatened. Whence not ineptly the Emperor Trajan called the spleen a tax. For just as with a growing spleen the remaining body will be weakened, so with an increasing tax, the people is impoverished."

[309] *De Partium Morbis,* in *Universa Medicina* (Paris, 1567), p. 298.

[310] *Fabrica,* p. 438: "A Paduan citizen, who had been detained three years in prison and finally died of jaundice, was employed for public dissection. We found the spleen to be very slight, and less thick and wide than others, with fat attached to its gibbous part, concreted in the form of very white and hard stone. The substance of the organ was dry and very hard."

[311] Benivieni appears to have made no such observation, although there is a possible reference to him on f. 20ᵛ.

cause of defect of the liver and of the stomach *as washhows to the kitchin;* thus in addition to defect it is lacking; *inferior kitchins need noe washhows.* Liver, smaller not bipartite, in faculty more precise but weaker; there the spleen necessary; where stronger and larger and the body uses more impure aliment there the liver *playes the splen too;* where *more fowle work cleane,* as in those of vitiated temperament; they employ vitiated and impure aliment. [In those with] larger spleen the *washhows exceed ye kitchin.*

In man a 2nd and 3rd [spleen] noted by Fallopius;[312] by Vesalius often in dogs;[313] on the other hand, sometimes lacking [according to Du] Laurens;[314] bold myrmidon of the Turks; when that WH saw scirrhus so that without undue damage was able to be removed *Mr. Gillow.*[315] On the other hand [according to] Galen, On the Use of the Parts, IV [4], Erasistratus persuaded by the opinion of the uninformed refutes his distinguished forerunners.[316]

Hence the spleen [is] less worthy, and nature degenerating into a monster easily errs here; thus it can be better to lack a spleen. Aretaeus, Book of Chronicles, I, 13, [states] that the liver is the balance to the spleen, but its faculty in health and in disease is by far greater. Galen describes this kind in his book The Use of Respiration.

Viscera of that sort in which [there is] blood; thus for the sake of blood and even more for the sake of veins than on the contrary for the sake of a vein. The liver [is] not the

35v

---

[312] Fallopius, f. 179ʳ: " . . . here in Padua I publicly dissected a cadaver in which I found a triple spleen—one of proper size, the second less than half, and the third about the size of a pigeon's egg. Each of these possessed its own veins, arteries and nerves, and the position of all was the left hypochondrium. The substance and covering of all were the same, and generally speaking everything was the same."

[313] *Fabrica,* p. 439: "In addition to the attachment of the spleen, with the aid of the omentum, to the back and stomach, from the peritoneum where it binds the diaphragm an inconstant number of thin fibers are sometimes attached to the gibbous part of the spleen, binding it to the diaphragm. But as fibers of this sort are very rarely found in man, so never in the dog."

[314] Du Laurens, p. 323: "In former years the body of a youth of good constitution was dissected in Paris, and found to be without a spleen."

[315] *Mr. Gillow* has not been traced.

[316] Erasistratus declared that the spleen serves no purpose and was a mistake of the Creator (*De Usu Partium,* IV, xv; Daremberg, I, 318–119).

origin of the veins, because the veins [develop] before the liver in the embryo;[317] in the egg.

Account from Bauhin,[318] see Annotations **WH** &c. &c.

Parts of the viscera: quantity, shape, substance   36r

Size   Liver with spleen because of balance; as much as by its nature it exceeds the [other] viscera, so much does it lack in flesh or fat, as I said. And as much as it exceeds in parts, so much in the use of them, wherefore the greater viscera concoct better; but [the smaller are] sluggish, gluttonous, pusillanimous; on the other hand, the muscular, brawny [are] brave according to the spirit.

Likewise a balance between the heart and liver, as between the liver and cerebrum; wherefore as much as [they lack] healthy and florid color, so much do they lack in operations of the spirit.

Wherefore disease of the liver [if] too small; of which the sign, according to Avicenna, [is] shortness of fingers;[319] thus size, *Jhon Bracey*,[320] huge, *as bigg as an ox liver; liver-grown*; very emaciated, curved because of weakness, dying from fistulas.

Man proportionately [has] large liver and spleen; large liver in embryo, but spleen smaller than liver; (Andreas Vesalius regarding candid youth of Diogenic ways hanged in Monselice).[321]

Large spleen in my sister[322] Δ 5 lb. and *Mrs. Yeoung*[323]

[317] Cf. n. 293.
[318] Bauhin, pp. 270–279.
[319] *Canon*, Lib. I, fen ii, doct. 3, cap. 1: " . . . just as shortness of the fingers which indicates smallness of the liver."
[320] Cf. f. 39ᵛ for another reference to *Jhon Bracey.*
[321] *Fabrica*, p. 438: "That one hanged at Monselice was brought to Padua for public dissection and displayed so large a spleen that it was scarcely smaller than the mass of his liver, and attached to the anterior part of the liver, it also extended to the anterior part of his stomach. The substance of the liver wholly resembled the substance in the organ of those who are healthy. He was a youth with a very fair and smooth skin and a nature far from melancholy; because of his Diogenic ways he had been pleasing to the people so that three time he avoided the gallows. First whipped in Venice, he came to Padua where he was deprived of his right eye and hand; again seized at Monselice—for he had once escaped from prison there—he performed the final act of his tragedy, and especially because of his ways he was employed for dissection by the students."
[322] This must be a reference to Harvey's sister Anne (d. 1645), or to his half-sister Gillian (d. 1639).
[323] *Mrs. Yeoung* has not yet been identified.

[extended] as far as pubic bone [and] counterfeited pregnancy; Colombo refers to 20 lb. in several;[324] wherefore some [look] gravid with a [very large] spleen.

36v
Shape as body permits, wherefore those in whom the venter long, [have] a longer spleen; [also] it receives its shape from the containing parts; because their operations do not depend on shape; [a particular] shape is not necessary.

The liver [is] branched in some, simple in others; in some as here, in others in a manner in between; [occurs] in different species as well as in the same species; in the sting ray [is] like Neptune's trident. In man simple, round and long, divided only by entrance of portal [vein] and by umbilicus.[325]

* Where little extensions, lobes in some; WH convex and *concave* like hoof of horse; concave *to imbrace the stomach*; branched [liver] seen in man; Fernel saw 5 lobes like the ancients.[327]

*from these obscure divisions John Caius contends on behalf of the liver[326]

[324] Colombo, p. 488: "In addition, spleens so large that each one of them by far exceeded a weight of twenty pounds."
[325] Here Harvey refers to the ligamentum teres, which connects the umbilicus to the under surface of the liver. Frequently, the anterior margin of the liver is notched by the ligamentum teres, which is the fibrosed remnant of the umbilical vein of the foetus.
[326] Here the reference could be either to Caius' notes to his edition of Galen's *Anatomical Procedures* in his *Libri Graeci* (Basel, 1544), or to the fuller statement referring to those notes in his *De Libris Propriis* (in *The Works of John Caius*, ed. Venn, [Cambridge, 1912], p. 82): "In the same commentaries I made a few remarks about the fibers of the liver, which in Greek are called lobes, because in Galen's time as well as today a controversy was directed against him on the number of lobes. Recent authors affirm that there are no lobes or only two lobes of the liver; the ancients wrote that there are two, three or four, according to the size of the liver. Graver and more learned men give their adherence to the latter and the less thoughtful and ignorant to the former. All error lies in the misunderstanding of the lobes. If those who relate that there are no lobes say that there are no divisions which make the lobes, they deny their perception, for the liver has been divided into two large parts in that region where the umbilical vein is introduced into the liver. Also it is divided in its cavity by certain divisions in the middle of which the parenchyma rises up into fibers, although not large and in considerable part completed from the remaining substance of the liver, as in the case of the lung where in Greek they are called lobes."
[327] *De Partium Corporis Humani Descriptione*, in *Universa Medicina* (Paris, 1567), p. 18. Galen described the liver in man as having five lobes, a description he borrowed from comparative anatomy. His teaching persisted into medieval anatomy, probably because no one thought of checking the

Galen recounts the separation of one lobe, the cause of jaundice and pain, On the Use of Parts, VI, 4:8, which restored, immediately the pain ceased. The followers of Avicenna have given 5 names: hearth, dining table, ploughshare, helmsman, &c.

Wherefore WH I believe there have been men in the past [with many lobes to the liver], as there are now rarely; and in part as Herophilus [states that it may have up to four lobes], G[alen], T[ome] 1, [col.] 281,[328] for the matter is so apparent that it is impossible to be deceived; and Aristotle says in the same species in some [it is] branched, in others simple as in the human species WH. Nevertheless Aristotle says round in men like that of the ox.[329]

The shape of the spleen [is] like the tongue of an ox [or] 37r the sole of the foot;[330] [it is] slightly bowed out on the left side; a little concave on the inner side, toward the stomach. Uneven surface, a little rough with some tubercles. In dogs *long and narrow*, rounder than in cloven-hoofed animals.[331] Longer than in the multifidous, as I said of dogs. Less distinct than in the solipedous, [being] partly long partly round.

Because humid and lax for receiving humors, in diseases [it has] a varying shape and in conditions contrary to nature [is] round, squarish. I have seen it large and filled with protuberances when dissected in wolves.

Its tunic, vessels, veins, arteries and nerves are divided

---

description against the actual organ in either a post-mortem examination or a formal dissection, or if they did they had no wish to question Galen's authority. Even Vesalius pictured a five-lobed liver in his *Tabulae*.

[328] Here the published transcription errs in giving G: T: 1581. The correct reference to volume and column permits us to determine that Harvey was using the *Opera Omnia* of Galen published by Frellon: Lyons, 1550. Vol. I, col. 281, contains part of the *De Anatomicis Administrandis*, bk. 6, ch. 8; Kühn, II, 570.

[329] *H.A.*, 496a:20: "The liver of a man is round-shaped, and resembles the same organ in the ox."

[330] Du Laurens, p. 321, cites Hippocrates, *De Corporum Resectione*.

[331] *P.A.*, 473b:30, 474a:1: "The spleen again varies in different animals. For in those that have horns and cloven hoofs, such as the goat, the sheep and the like, it is of rounded form." " ... On the other hand, it is elongated, in all polydactylous animals. Such, for instance. ... in man, and in the dog."

by parenchyma, blood and spirits. Very thin tunic and membranes from the peritoneum; spleen from the omentum.

Blood and spirits as the contained, and other things containing. Contrary to [the opinion of] Aristotle[332] the blood is contained in the liver [and not in the spleen] because there is no anastomosis. Whether or not it be contrary to Bauhin ['s opinion], yet it does not agree with his use.[333]

2 From WH, regarding the structure and course of the veins, for when the portal has entered [the liver] here, it is divided into branches, they into others, always in the hollow of the liver[334] as far as its knob. But, on the contrary, the [vena] cava [enters] into the gibbous parts;[335] the vessels are joined here like the fingers of both hands. Branches of the portal very rarely to the gibbous part, as on the contrary those of the [vena] cava to the hollow [portion]. Wherefore, as WH, many seek anastomosis but do not find it;[336] who find, like Bauhin, that only one is demonstrated.[337]

The portal vessels are different [for] they send quasi 37v branches into the parenchyma. WH branches of the [vena] cava [are] in the form of a cradle; the follicle in which obstructions of the liver occur [according to] Hippocrates.

All the veins of the spleen, as of the liver, have a thinner

[332] *P.A.*, 669b: 15.
[333] Cf. n. 297.
[334] The portal vein enters the liver at the porta hepatis, on its concave, inferior surface.
[335] The curved posterior surface of the liver. The hepatic veins join the inferior vena cava by a number of openings.
[336] The hepatic veins begin in the substance of the liver as the central vein of each histological lobule. Into these drain blood from the liver sinusoids composed of fine capillaries from the hepatic artery and portal vein. This would therefore be an anastomosis, but it would not have been visible to Harvey. He is referring to a possible connection between the portal vein and inferior vena cava which does not exist as such in the adult, although in the foetus the two are connected by the ductus venosus. The obliterated ductus venosus remains in the adult as the ligamentum venosum.
[337] Bauhin, p. 288: "The superior [branches] little by little are collected into larger and these into others until they are gathered into two very large branches in the anterior site of the cava where it is attached to the liver and rests on the diaphragm; extending, they form one trunk; hence the portal vein is commonly said to arise from the hollow of the liver, the cava from the gibbous part."

101

tunic, **WH** because issuing forth they borrow another tunic from the peritoneum or pleura.[338] As the liver [has] many branches of veins so the spleen of arteries which [go to] the liver only in the hollow part; use of the arteries [is] to concoct crude aliment, to agitate excrement and carry it down to the kidneys; wherefore splenetic vessels expurgate through the kidneys as I myself [have shown].

Both [receive] slender nervules from the 6th pair near the roots of the ribs,[339] and the liver [has] another from a branch carried to the orifice of the stomach and thence it enters with the portal [vein]; distributed into the tunics, wherefore sensitivity only in the tunic, and blunted and slight, [indeed] scarcely any, in the parenchyma.

Whether there are spirits in the liver or whether they are made [there]. **WH** I [believe there are] none separate except air and a portion of the blood, wherefore arrangement of the blood just as of cold water, and to dispute about the spirits is [to dispute] formally about the blood or about innate heat.

The substance [of] the flesh [is] a spread-out paren- 38r chyma, as if effused, extra-venate, concreted blood. Since the veins must be distributed into little branches for the sake of coction, [they are] *Twined and intertangled* in the soft parenchyma lest they be injured; [the parenchyma] separates, supports, spreads, and, by its mild fostering heat, promotes coction. Like the pancreas and glands for the larger branches, so the parenchyma for the smaller; wherefore [according to] Aristotle [it is] like an anchor,[340]

[338] The portal vein is posterior to the peritoneum except in its course from the stomach to the porta hepatis where it is between the two layers of the gastro-hepatic omentum. Similarly, the splenic vein on emerging from the spleen lies in the lieno-renal ligament before going behind the peritoneum. However, it is not correct to say that these veins get a true tunic to these walls. It is also not correct to say that the inferior vena cava gets a tunic from the pleura, for after passing through the diaphragm, it enters the right atrium of the heart. Here it is related to the pericardium, not to the pleura.

[339] The sixth pair of cranial nerves in the early seventeenth century consisted of the modern glossopharyngeal, vagus, and accessory, together with the sympathetic trunk. Here Harvey refers to the vagus and probably the abdominal portion of the sympathetic chain.

[340] Aristotle believed the viscera anchored the great vessels to the posterior abdominal wall by means of their branches to these viscera. This is

as the placenta for the umbilical vessels. WH As *cotton to kepe a Jewell*. Wherefore [it is] for the sake of the veins rather than of the veins for the sake of the viscera, for all things for the sake of native heat and blood. It differs from flesh as cartilage from bone or *prowde flesh*.

[The substance] of the spleen [is] as much rarer than [that of] the liver,[341] as the lung is thicker. Therefore it admits thick juices which the more compact structure of the liver could not receive; WH rarer so that the many arteries may pulsate.

Sanguineous color of the viscera wherefore the argument of sanguification, and [according to Du] Laurens coloration of blood from the liver[342] X..1, blood of same color[343] in the mesenterics. 2, somewhat pallid color of liver in fish, in marsh fowl. *Goose durty color; flownders* [have] a liver variegated in color and consistency, in some pale in others very whitish, *as in a goodens season; so powltry fatt white liver;* sting ray *light yellow,* other colors in crustacea, fish. WH I believe *kingman* between *pale yeallow and light red,* nevertheless in all darker blood; in a healthy man *sumwhat russet-Iron grey.* Color of the viscera according to the blood. *Splen* brighter in color than the liver; *Catt weesel.*

Aristotle in regard to vitiated blood; darker whether hot or cold, wherefore fleshes and viscera of darker color in man; spleen especially, for [when] dilated it attracts dispersed and vitiated blood; wherefore in a man of vitiated temperament in whom all the blood [has been] vitiated, the liver [is] similar to the spleen. On the other hand, in those of the best blood, in the same way the spleen brighter, of the same color as the liver.

---

expressed in his *P.A.*, 670a:15: "The great vessels send such branches to the liver and spleen; and these viscera—the liver and spleen on either side with the kidneys behind—attach the great vessels to the body with the firmness of nails." Oddly enough Aristotle goes on to say that the aorta does not send a branch to either the liver or the spleen, which rather spoils his argument.

[341] The splenic pulp is somewhat softer than that of the liver.

[342] Du Laurens, p. 311: " . . . that by its special and innate force [the liver] adorns the blood . . . with its red [color]."

[343] Here the published transcription erroneously gives *equae coloratus,* rather than *aequae coloratus.*

Approbation of Aristotle; wherefore in splenetics [there 38v
is] darker blood, [hence] black scars of which the flesh is
[dark] like Quiloa earth[344] or guinea fowl; perhaps the
same cause of the spleen. Yet they affirm there is a tasti-
ness, but perhaps hunger and price responsible.

Temperament of the viscera [is] hot and humid, *yett less
hott than ye blood*, and the liver *less than ye splen* because of
the arteries. [According to] Galen as much as it is more
humid than the skin (according to the rule) so much is it
softer, and so much the hotter as its supply of blood is
greater.[345]

Passions of the viscera. I shall refer to what I have seen. 39r
Just as the temperament is softer or harder according to the
humidity, so larger or smaller veins relative to the heat.
The *Splen* in cachetics or scorbutics [is] in some (not in all)
very fully stuffed with blood so that [it is like a] *Bagg of
blood;* usually in those in whom the body is most transpir-
able, fleshy and florid, or a fatty, small spleen.

On the other hand, in those of lean, melancholic complexion, with yellowish or obscure color, I have seen
greater laxity; and now as far as the pubis I have *easyly
taken vp*, and Colombo, as I said, [saw a spleen weighing]
20 lb.[346]

In some pale cachetics [it is] black and livid; outwardly
watchet;[347] also dull green; tunic with cartilages like *finger
Nayles*[348] or *horne softned in water* on testimony of *Dr.
Flud;*[349] *Mr. Waker;*[350] also Bauhin[351] and Colombo[352] refer

[344] Quiloa is an island off the east coast of Africa, about 150 miles south
of Dar es Salaam, held by the Arabs throughout the Middle Ages and under
Portuguese rule in the sixteenth and seventeenth centuries. The several
kinds of "earths" given in the seventeenth century pharmacopoeias are
named after their places of origin. "Quiloa earth" has not been found in the
contemporary literature. Judging from Harvey's simile it must have been a
dark reddish, purple substance.
[345] *De Usu Partium*, IV, xv; Daremberg, I, 320-321.
[346] Cf. f. 36r.
[347] A late Middle English word meaning "of a light blue or sky blue
color."
[348] The published transcription incorrectly gives "fungi *Nayles.*"
[349] Robert Fludd (1574-1637), M.D. (Oxon.), 1605, Fellow of the College of Physicians, 1609, and Censor, 1618. His book *Medicina Catholica*
(1629) contains the first public acknowledgement of the correctness of
Harvey's views on the circulation.
[350] The only man of this name traced is Richard Waker, of St. Michaels,

to having seen such; according to early writers a hot and humid spleen, according to these later ones hot or cold and dry; wherefore I believe hard in some because of thick nutriment; in others cartilaginous through increased heat.

WH I have seen abscesses and putridity in all the [other] parts but have not seen or read of it in the substance of the spleen.

Liver —
```
         │ 1
   small │ 2
         │ 6
      4  │      │ 10
      7  │      │ 11
      8  │likewise│ 12
         │      │ 13³⁵⁵
         │ 2
         │ 3
   large │ 5
         │ 9
```

I have seen the liver *russet hard contracted* without blood, *yeallow cankret substance* in the veins; 1, from a fever in the whole obese body [of] *Sr Rigdon*;³⁶³ 2, swollen with the gall bladder more distended so that [it was] filled with urine³⁶⁴ and stones, both from jaundice. WH In which case I believe that the calculi [arose] more from the jaundice from long [enduring], very hot, concreting material than, on the contrary, the jaundice from the calculi.

2, Likewise *russetish* huge and hard; plainly a hard tumor, almost without blood, rough surface.

3, Between *Russet and purple* extended *as bigg as an ox liver Jhon Bracey.*³⁵⁶

4, *Beginning to be discolored Joan Jhonson* dead from malignant fever and hemorrhage after a quartan.

5, Inflamed, distended, full of black blood.

6, *Palish durty color* with mucous *on ye coat;* dropsical, sweaty matter, *little blathers* or like *blisters.*

7, Hard from tension and putrescent through 6 or 8 months; in a fever with turbid urine, *like a heape of pus* of

---

Cornhill, an upholsterer, who married Alice Grimsdich at All Hallows on February 17, 1599. Cf. J. Foster, *London Marriage Licenses. 1521–1869* (1887), col. 1396.

³⁶¹ Bauhin, p. 271: "In those in whom the spleen is preternatural it exhibits such color as the humor provides, livid, leaden, leek-green; sometimes covered by a tunic like cartilage in color, thickness and hardness, but in the dog a brighter red than the liver.

³⁶² A reference to the very heavy spleens, cf. f. 36ʳ. Colombo, p. 488; "which cartilage covered externally."

³⁶³ Cf. n. 170.

³⁶⁴ Here Harvey refers to the bile as urine. He uses this term again, e.g., f. 41ᵛ.

³⁶⁵ Harvey appears to be grouping the numbers of his pathological findings, which are continued on f. 40ʳ.

³⁶⁶ Cf. f. 36ʳ for another reference to *John Bracey.*

pale yellow color, *noe shape or particle remayninge* of the liver in that one who had a cartilaginous spleen. *Mr. Benton* who I hoped healthy after prognostication.

8, Usually in dropsy or otherwise where the liver *russetish splen watchet greenish or lead color* if sick of cachexia.

9, Huge aposteme[357] through many months from very fetid pus, *2 or 3 gallons* and water with viscous convoluted pannicles *as glew stepened in water, or Isonglass;* returned to hospital.

10, Dead woman; inflammation of the liver *of a botch* in the tunic; *red, hard,* cancerous.

11, Roughness of the liver: scabby *hard soakt* in water. 40r

12, Small abscesses as in rabbits; *rotten sheep.* Abscess *like an eagle stone,*[358] seal; *East Indy Bezar*[359] *in the liver of a goate.* Plater recounts [a case] from *a* fungous *substance,*[360] such as I have not seen. I observed these things in the hospital, as well as in the hospitals of Italy, with much nausea, loathing and foetor. I have forgotten many things.

Conclusion: See with me some time.

None from chronic diseases in which there is neither

[357] An abscess.

[358] A mass of argillaceous oxide of iron, often hollow in the center and having a loose nucleus. This apparently solid stone, which rattled when shaken, was, according to fable, found in the eagle's nest; hence its popular name. The Greeks believed the eagle could not lay her eggs unless the stone was in her nest. In folk medicine it was thought that an eagle's stone bound to the thigh of a parturient woman would facilitate birth or inhibit abortion. Worn suspended from a necklace, it was thought to ward off epilepsy; cf. W. G. A. Robertson, "The use of the unicorn's horn, coral and stones in medicine," *Ann. Med. Hist.* (1926), 8:240–248. Sir Thomas Browne, in *Pseudodoxia Epidemica* (1646), bk. II, ch. 5, says briefly: "Physitions joyn it [the loadstone] with Aetites or the Eagle stone, and promise therein a vertue against abortion." In the 1658 edition Browne has his doubts about the efficacy of the stone when he says: "Whether the Aetites or aegle-stone hath that eminent property to promote delivery or restrain abortion, respectively applying to lower or upward parts of the body, we shall not discourage practice by our question."

[360] Bezoar stones are concretions found in the stomach of goats and deer. They formerly had the virtue of being regarded as a highly potent antidote to poison. They were, of course, quite useless. The oriental bezoar stone was found in the stomach of a goat; it is small and has a smooth surface. The occidental variety came from the stomach of the various members of the deer family; it is larger than the oriental and has a rough surface. Genuine ones were naturally rare; this obviously increased their efficacy against poisons in the minds of their credulous but proud owners.

[360] Plater, *Praxeos* (Basel, 1625), pp. 442 ff.

large abscess nor a corruption of some sort in the viscera as *liver lungs*[361] *Brayne kidney stomach* and those things totally; and that these things observed [are] rather the fruit of diseases and cause of death, than the causes of diseases; and the fruit the effect of; and that 1st cacochymia[362] occurs from errors in diet and by weakness of nature; from cacochymia [also arises] poor bodily nature. From this [one gets] corruption of some of the viscera and thence death [when] the viscera [are] totally destroyed. For it is astonishing that in the smallest part or none, that with the whole liver corrupted, [persons] have lived Δ for months. [According to] Riolan filled with gypsum-like pituita and contracted into a ball.[363]

Gall [bladder]

[Shape] of a very long pear compressed from base into neck; twin veins in cysts by which it is nourished; wherefore X it is nourished by bile; because it does not respire, for arteries here [come] from the coeliac. Slender nerves from branch as if from the liver; [the gall bladder] *darke yeallow;* some *black,* some *rusty;* likewise *greenish blewish,* wherefore in different ones bile yellowish, that of calves rusty-colored. Black, leek-green or green.

Blowe vp
Shape
Vessels

Color

In some large in others small according as the liver of worse or better color and consistency. In fish large; although bile is from heat it occurs in fish with cold [temperament]; animals, little [choleric] blood since in horses *Chamel Elephant stagg* none;[364] *nay* in the seal, dolphin, among the marine, very hot.

Size

Generation

[It is] clear that it occurs from heat; for by cold it lique-

[361] The published transcription erroneously gives "*longs.*"
[362] A depraved state of the humors.
[363] Riolan, p. 209.
[364] *P.A.,* 676b:25: "For in some animals there is absolutely no gall-bladder at all—in the horse, for instance, the mule, the ass, the deer and the roe; and in others, as the camel, there is no distinct bladder, but merely small vessels of a biliary character. Again, there is no such organ in the seal, nor, of purely sea-animals, in the dolphin." Harvey, like Aristotle, is wrong when he says the seal has no gall bladder. As Ogle points out in the notes to *P.A.,* the *Phoca vitulina* has a gall bladder.

107

Temperament

fies; [according to] Aristotle;[365] like sperm. But the humor not so hot as jaundice; in the torpid, dull, inactive, slothful, somnolent, all of which from cold; WH by this origin a contrary condition; *Will Halte flash and flame*; not however rage as this is the seat of anger; but I believe it is oppressed by heat; as a supply of aliment and to that degree cold.

Division into

1, tunics, afterward; branches into liver, gall duct, gall passage the same for egress and ingress; 2, *partes Conteyning:* bladder; 3, contained, afterward; sinus discovered by Jasolino Hipponiata.[366]

1st the nerves display themselves from the orifice of the stomach. Branch [of] 6th [pair] above;

Principal part, gall duct, and X bladder, for where the bladder is lacking the duct [is present] as in horse, mule, ass, camel, elephant, thrush, pigeon, seal, dolphin, sea calf, *stagg deere*,[367] viz., in which the liver can best be erected; lamprey, whale, orc, *Oestridg*, the duct only.

2nd by a branch of the nerves into the jejunum and gall [bladder][368]

Wherefore when the liver healthy and the blood of very 411 sweet juice, there the [gall] bladder is lacking.[369] 1, thus in fish when the liver [is] *discolored*, impure; a large supply of gall, for the voracious enjoy impure aliment. 2, in men *the blacker the liver the greater the gaule.* 3, wherefore in antiquity

3rd artery extended here 4th under it by a quasi duct 5th portal vein above artery;[370] *nerve and vein at same time all things in omentum*

[it was thought that] the length of life [was related to] the absence of gall as in stag and solipedous,[371] *grus.* 4, other

[366] *P.A.*, 649b:30: "We can therefore understand how some substances are hot and fluid so long as they remain in the living body, but become perceptibly cold and coagulate so soon as they are separated from it; while others are hot and consistent while in the body, but when withdrawn undergo a change to the opposite condition, and become cold and fluid. Of the former blood is an example, of the latter bile; for while blood solidifies when thus separated, yellow bile under the same circumstances becomes more fluid."

[366] Giulio Jasolino. *De Poris Colidochis & Vesica Fellea.* Naples, 1577.

[367] The bile ducts are, of course, present in those animals that lack a gall bladder.

[368] In this paragraph Harvey is referring to the branches of the vagus nerves.

[369] *P.A.*, 677a:19: "All animals, then, whose liver is healthy in composition and supplied with none but sweet blood, are either entirely without a gall bladder on the organ, or have merely small bile-containing vessels; or are some with and some without such parts."

[370] In the gastro-hepatic omentum the portal vein is placed behind the hepatic artery. The common bile duct is also in front of the portal vein on the right-hand side of the artery.

[371] *P.A.*, 677a:30: " . . . deer and animals with solid hoofs. For such have no gall bladder and live long."

animals partly have partly have not gall, as in *ratts mise* according to Aristotle[372] [IV], p. 506.

Necessity of the
bladder Δ that
in the embryo
not empty in the
dead through
synathresis[375]

duct

Likewise it has happened in man; in sheep and goats; [according to] Aristotle those on Naxos [used more for] magic rather than [for] augury;[373] in Chalcis, Euboea, *None*.[374] WH Wherefore prophecy from the gall in the victims. Hence it appears from all these things that the bladder for the sake of the duct, and not necessary, but in those in whom an abundant supply of bile; wherefore nature made the duct for expurgating the liver, which [is] in many branches in the liver [near] the hollow where the portal vein begins to ramify.[376] Thence to the intestines.[377] NB Nothing more difficult in the entire dissection of anatomy. [According to] Vesalius it arises from the liver.[378] It is terminated by many branches in the intestines,[379] *sometime Higher some time lower*. Wherefore diversity of authors over the termination of the duodenum [and] the origin of the jejunum. WH In the *Goose* [the termination is] uncer-

---

[372] *P.A.*, 676b:29: "Even within the limits of the same genus, some animals appear to have and others to be without [the gall bladder]. Such, for instance, is the case with mice."

[373] *P.A.*, 676b:35: "The same has occurred in the case of sheep and goats. For these animals usually have a gall bladder; but, while in some localities it is so enormously big as to appear a monstrosity, as is the case in Naxos, in others it is altogether wanting, as is the case in a certain district belonging to the inhabitants of Chalcis in Euboea."

[374] The published transcription gives the word *Nomen*, which is meaningless in the context. More likely, Harvey's word is *"None."* *H.A.*, 496b:25: "And, by the way, the absence above referred to of a gall-bladder is at times met with in the practice of augury. For instance, in a certain district of the Chalcidic settlement in Euboea the sheep are devoid of gall-bladders, and in Naxos nearly all the quadrupeds have one so large that foreigners when they offer sacrifice with such victims are bewildered with fright, under the impression that the phenomenon is not due to natural causes, but bodes some mischief to the invividual offerers of the sacrifice."

[375] Here the printed transcription gives the meaningless *sunatharsin*, although the likely word is *synathresin*.

[376] The hepatic ducts emerge from the liver at the porta hepatis in front of the point where the portal vein enters.

[377] The common bile duct opens into the second part of the duodenum some 8 cm. from the pylorus.

[378] *Fabrica*, p. 435.

[379] This statement would appear to apply to the common bile duct, in which case Harvey is confused, for the bile duct is almost always single and has one opening in the duodenum. Near the duodenum it is joined by the pancreatic duct. In rare cases an accessory hepatic duct may run alongside the bile duct to terminate separately into the bowel.

tain for [it has] a convoluted intestine as well as an expanded liver.

In the night owl after great fold inserted into intestines
It is inserted between two tunics and runs obliquely. Sometimes by branch to the stomach, wherefore by the bitter bile of the anger of families [according to] Bauhin.[380] Nevertheless [according to] Vesalius.[381]

Vesalius regarding a sailor not vomiting bile,[382] WH but 41v perhaps because exit within the tunics in the intestines.

Vse Bile [flows] from the liver into the intestines; where [one gets] jaundice and white faeces[383] from an obstruction. As sometimes the intestines [are] distended by flatus, therefore [one gets] jaundice from colic, and [in] colic, urines[384] are restrained because [the bile duct] is easily compressed and obstructed by an oblique insertion.

Uses Bile by descent stimulates excretion. It lubricates the in-intestines and drives out the chyle and tints the faeces;

[380] Cf. n. 382. Bauhin, p. 298: "The other passage, called the bile duct, supported by the inferior membrane of the omentum, draws thick and mixed bilious excrement from the liver."

[381] *Fabrica*, p. 436: "But a portion of this passage which we are presently considering is not always extended into the stomach, because in dissection when it has been freed from the surrounding vessels and membranes, no opening appears by which a branch of that passage might be led into the stomach."

[382] As late as the opening of the sixteenth century, Gabriel Zerbi maintained that two ducts always issue from the gall bladder, one emptying into the duodenum and the other into the fundus of the stomach. In conjunction with consideration of these matters, as well as his contention that the bile duct "is separated farther from the stomach's orifice than other anatomists generally realize," Vesalius described the following anomaly: "I never saw the smallest portion of the passage of the gall bladder extended into the stomach except in one of the oarsmen of a papal trireme. For as certain other of his organs, especially about the ribs and thoracic muscles, differed from the structure in other men, so this passage was divided into two branches, of which the lesser or more slender was inserted into the fundus of the stomach where first the vein which goes about the right side of its lower region joins to the stomach. He seemed to be of a hot and dry temperment. In order to learn whether he was accustomed to be bothered by bilious vomiting, as Galen asserts is frequent in men of this type, I carefully enquired of his friends if they had observed anything of this sort. They replied that they had never seen him vomit or be sick even in the roughest weather, and they all affirmed that he was very robust, that he could blow a horn with a very long breath and could by far surpass all his comrades in the depth of his voice and in counting numbers with one breath" (*Fabrica*, p. 436).

[383] Jaundice and clay-colored stools are two classical signs of obstruction to the common bile duct.

[384] Here Harvey means bile, as he does on f. 39ᵛ.

110

wherefore by jaundice the white and restrained urines,[385] and as much as they are more restrained so they are whiter from the obstruction.

It drives out and evokes worms in Δ horses, fish, children

In antiquity it was believed [to be] for stimulating or moderating the quality of the soul in the liver for the sake of sensation;[386] wherefore anger &c. The passage of the gall *as a branch torning back;* for the duct passes through and does not enter into; wherefore the bladder does not draw from the liver, but either voluntarily as [according to] Aristotle,[387] or by expulsion from the liver. Hence how *idle* [Du] Laurens disputes of those things[388] which the ancients and, on the other hand, Fallopius more mordantly;[389] by such valid arguments as rams: **WH** *Rams are but sheepe* their inflation not to be trusted.

WH I saw in a canine foetus ij ducts, one receded into the bladder. The passage of the gall *from* near the intestine rearward in the same tunic with the duct; this is not perceived unless *bearing the probe this way,* and interstice in place of a valve; use: it is regurgitated through the passage into the bladder just as into a diverticulum; because the transit is impeded at the intestines;[390] wherefore in all those abounding in bile the duct is like a bladder.

[385] Possibly Harvey means the contents of a gall bladder which follow impaction of a stone in the cystic duct with formation of a mucocele.

[386] *P.A.,* 676b:22: "There are therefore no good grounds for the view entertained by some writers, that the gall exists for the sake of some sensory action. For they say that its use is to affect that part of the soul which is lodged in the neighborhood of the liver, vexing this part when it is congealed, and restoring it to cheerfulness when it again flows free."

[387] *P.A.,* 676b:25: "But, when animals are formed of blood less pure in composition, the bile serves for the excrement of its impure residue. For the very meaning of excrement is that it is the opposite of nutriment. . . . the bile, which is bitter, cannot have any useful end, but must simply be a purifying excretion."

[388] Du Laurens, p. 243.

[389] Fallopius, ff. 178ᵛ-79ʳ: "In this connection I think I ought finally add that I have observed, two or three times at the most, that the passage which discharges the bile when near the duodenum divides into two canals, each of which, a little space intervening, is inserted into the same intestine and in each is preserved that skillful insertion which has been described. However, I have never seen any passage carried down to the stomach, despite the very weighty testimony of the ancients and of the divine Vesalius who saw this arrangement and attested the man in whom he saw it, although I do not deny that this is possible."

[390] Here Harvey refers to the sphincter at the termination of the common

Thus the gall [bladder] in some [is] near the liver, in
others near the intestines, so that in fish at the intestines
[and] in crow, quail, sparrow, swallow, bordering on the
whole intestine [according to] Aristotle, [De] Anima.[391]
WH On the other side away from the liver in serpents; X in
vipers, eels, serpent. In others partly in the liver partly in
the stomachs, as [according to] Aristotle in the horned
owl,[392] *Barble*. Some partly in the stomachs partly in the intestines as in hawk, kite; WH black duck, starling. Some in
a middle position either in the manner of the liver or of the
intestines as in the conger eel.[393] WH Δ in marine frogs in
either place between the liver and the stomach or at the intestines;[394] *nerer and farther from the liver*.

Valves formed in the duct at the orifice; ij or iij valves.

Slender passage through the tunics; some as Giulio Jasolino and Bauhin[395] (that they might preserve Galen's position that the [gall] bladder attracts from the liver), wherefore according to them pure *bile* through the duct through a
mixture; but X those in whom the [gall] bladder is not at
the liver, and thereupon in others unless for the reason that
the bladder in some may be near the liver; for why *thus re-
coyled to the liver* I do not see, and why held back (unless it
aids the coction by heat) which is seen as useless, for in the
embryo it is where they say [there is] no general coction.
Nor is it hindered at the intestines contrary to Fallopius.[396]

---

bile duct. It is now called the sphincter of Oddi, but was originally described
by Francis Glisson (1597–1677).

[391] *H.A.*, 506b:19: "The case is much the same with birds: that is, some
have the gall-bladder close to the stomach, and others close to the gut, as
the pigeon, the raven, the quail, the swallow and the sparrow."

[392] *H.A.*, 506b:23: "Others have it near at once to the liver and to the
stomach as the aegocephalus [perhaps horned owl]; others have it near at
once to the liver and the gut as the falcon and the kite."

[393] *H.A.*, 506b:17: "Other animals of the same species show the diversity of position; as, for instance, some congers are found with the organ
attached close to the liver, and others with it detached from and below it."

[394] *H.A.*, 506b:15: "Others have the organ in the region of the gut; in
some cases far off, in others near; as the fishing-frog. . . . "

[395] Jasolino, *De Poris Colidochis & Vesica Fellea* (Naples, 1577), p. 20;
Bauhin, p. 299.

[396] Fallopius, ff. 177[r,v]: "Hence nature, mistress of all the most exquisite
inventions, constructed the insertion of the passage through which bile is
carried to the intestines so cleverly that when the intestine is dilated, its

The tunic [has] ij membranes, one from the membrane of the liver, from the peritoneum. The other special [tunic] given fibers of all kinds, internally straight, externally transverse, fewer oblique in the middle; I have already spoken of the fibers WH.

The gall is contained, which they say is created for the 42v second concoction. WH Nevertheless gall [is present] in embryos where [there is] no second coction. Often calculi [there] just as also in the duct [according to] Fallopius.[397] WH See the black tophi[398] in a long jaundice *Dr. Gulstone grey like Bezar;*[399] very often a black concretion caused by jaundice; wherefore (as the Arabs note) in the beginning violent purgations make the jaundice incurable. For they putrefy the thickened bile and concrete it into tophi; wherefore obstruction more by impact; *partes cling together.* Incurable.

Crust in the interior tunic like [that of] the stomach lest   Crust it be harmed by sharp bile; wherefore it is not nourished by bile, for by fortifying opposes aliment X.

Kidneys, NEPHROI, as it were, urinators. *Reynes kid-* 43r *neys.* Action [is] *to draw away and convey oute of ye veynes by the vreters into the blather the* serous excrement; wherefore *seated to the great veynes and Arteries* and connected through veins and arteries to the emulgents.

---

tunic constricts the tunic of the duct and excrement cannot be emptied into the duodenum."

[397] Fallopius, f. 178ᵛ: " . . . the very numerous and very large stones which I have frequently found in the gall bladder, as well as in the wide passage from the liver to the intestines, but the space cannot be granted here [for further discussion]."

[398] A tophus (lit., "a porous stone"), a term applied to a hard swelling or node, and especially to the deposits found in the tissues around the joints in gout; often incorrectly called "chalkstones."

[399] Theodore Goulston (1572–1632), M.D. (Oxon.), 1610, and Fellow of the College of Physicians, 1611. He was Harvey's neighbor and founded the Gulstonian Lectures at the Royal College of Physicians. Gallstones are occasionally passed *per anum,* and in view of Harvey's reference to *Bezar* (cf. n. 359), such must have occurred in the case of Dr. Goulston.

113

In those in which that serous [excrement] does not abound as in fish and in all venomous serpents and oviparous quadrupeds because [there are] bladders in place of lungs. They do not drink, or little or rarely; and if drink, not drink for the sake of food. *sipp-sipp;* wherefore they do not have a serous superfluity.

In which not

Therefore these [animals] are lacking kidneys, ureter, bladder and where the bladder and kidneys are lacking, whether on the other hand [present] in birds which have a caruncle at the back.[400]

But in those in which the kidneys and bladder are lacking X in the tortoise and in *an efft* [newt] a bladder is found WH without kidneys.

Likewise in those in which that excrement is consumed in other uses as for scales in fish and in birds and in the testacea, *fethers, scales and shells.*[401]

---

[400] Harvey is probably referring here to the flattened kidney in birds, which is lobular and placed alongside the vertebrae.

[401] This section, on those animals which have kidneys and those which do not, is based on Aristotle and contains a number of errors. Aristotle did not observe the kidneys of Ovipara, and his statement that they are absent has been blindly accepted by Harvey. However, Aristotle stated that the tortoises were an exception, but Harvey wrongly says they lack kidneys and bladder. Aristotle's views are given with dogmatism in his *P.A.*, 671a: 5–30: "Those animals on the other hand, that are without a lung of this character, and that either drink but sparingly owing to their lung being of a spongy texture, or never imbibe fluid at all for drinking's sake but only as nutriment, insects for instance and fishes, and that are moreover clad with feathers or scales or scaly plates—all these animals, owing to the small amount of fluid which they imbibe, and owing also to such residue as there may be being converted into feathers and the like, are invariably without a bladder. The tortoises, which are comprised among animals with scaly plates, form the only exception; and this is merely due to the imperfect development of their natural conformation; the explanation of the matter being that in the sea-tortoises the lung is flesh-like and contains blood, resembling the lung of the ox, and that in the land-tortoises it is of disproportionately large size. Moreover, inasmuch as the covering which invests them is dense and shell-like, so that the moisture cannot exhale through the porous flesh, as it does in birds and in snakes and other animals with scaly plates, such an amount of secretion is formed that some special part is required to

What follows

Likewise in those in which there is not a bladder [they excrete] by spitting out. *sipp sipp.* and birds of prey, not entirely; the eagle Δ WH and Aristotle jeers at Hesiod because he introduced a winged cupbearer for Jove.[402]

Those in which there are

Likewise in more perfect and hotter animals [there are] both kidneys and bladder and because hotter perhaps difficult to understand.

But from anatomy those in which lungs especially full of blood and a supply of blood; wherefore they require most refrigeration, and the most thirst; and they drink, wherefore they are abundant in serum [and] the excrement [is] of the second coction.

And so in them these organs of evacuation [occur]: kidneys, bladder.

And so in all animals which generate more perfect and hotter animals than those which [are] from an egg or seed, and these drink much.

Men who sweat when they drink much abound in urine; drunk to bed piss free; wherefore from errors in diet judgment of urine [can be made] and without drink in winter time [more urine than in summer];[403] those of sedentary

---

receive and hold it. This then is the reason why these animals alone of their kind, have a bladder, the sea-tortoises a large one, the land-tortoises an extremely small one.

"What has been said of the bladder is equally true of the kidneys. For these also are wanting in all animals that are clad with feathers or with scales or with scale-like plates; the sea and land tortoises forming the only exception."

[402] This is from Aristotle, *H.A.*, 601a:30: "Birds of prey, as has been already stated, may in a general way be said never to drink at all, though Hesiod appears to have been ignorant of the fact, for in his story about the siege of Ninus he represents the eagle that presided over the auguries as in the act of drinking." The translator notes that the reference to Hesiod is unknown.

[403] Crooke, p. 189.

115

life who never sweat and [have] humid bodies; or by a humid diet and from morbid liquefaction as dropsy [resort to] the urinal. Because from contraries much humidity is consumed through sweat or [excreted] insensibly. Thus with the bowel constricted they excrete much urine. Wherefore the kidneys have foul odor and taste; unless in the delicate and fat.

Another use of the kidneys as of other viscera, that they be an anchor for the veins[404] lest they be compressed; wherefore according to Galen,[405] tension of the veins and arteries from inflammation of the viscera; thus hardness of the pulse.

Another use depends upon their fleshiness [when used] for coction; wherefore in those in which much bladder fleshy portion near the very large vein, as in fish and birds, in the cavity of the coxendix. Whether kidneys WH doubts. *In an Eft* [newt]; fleshy in calf at loins. WH Whether in place of kidneys because it is a bladder.

Different opinions on how the kidneys expel urine.[406] Some [maintain it is done] by attraction, others by expulsion of the veins. Some [that it occurs] by a succession by which exit is given [according to] Erasistratus.[407] Spontaneously [according to] Averroes[408] and Aristotle,[409] Tom.

[404] Cf. n. 340.
[405] *De Locis Affectis*, II, 3; Daremberg, II, 509.
[406] The physiology of the kidneys was ill-understood in the early seventeenth century and before, many theories being advanced for the production of urine. Harvey's account which follows is vague and merely repeats the thoughts of his predecessors and contemporaries. It was to be left to Lorenzo Bellini (1643–1704) to give the classical account of the structure of the kidney in his *De Structura et Usu Renum* (Florence, 1662). In this he described the excretory ducts and postulated a physical theory for the secretion of urine.
[407] The view is expressed in Du Laurens, p. 254, that Erasistratus considered the secretion of urine as the result of its filling a vacuum, presumaby created in the bladder by issue of the urine previously contained. This is in accordance with Erasistratus's idea of *horror vacui*, and Harvey very likely employed Du Laurens' interpretation as his source, since the original statement, in Galen, *De Facultatibus Naturalibus* (I, 16), is not so precise.
[408] It seems quite likely that Harvey once again has referred to the convenient Du Laurens, p. 254, who remarks of Averroes that it was his view that nourishment is not attracted by the parts, but is carried to them by its own movement, for by concoction nourishment takes a new form with which it acquires the faculty of moving itself.
[409] According to Aristotle, *P.A.*, 670a:15, the kidneys were constructed

116

4: 666 Δ fruit of trees. All these things are treated skilfully by [Du] Laurens,[410] but peace to that great man, WH I do not agree. He received it from Hippocrates through his teacher Duretus;[411] he says that urine contains three materials: the matter of drink, ichor, liquid from blood; 1st that all a critical evacuation by expulsion; 2nd by attraction by which exit is given for the collected liquid. WH The difference of the matter of drink from ichor; both excrement of the second concoction.

The question properly is not of the excretion of urine 44r but of all excrement and all motion in the body; and in the same way urine resembles the rest through its own ducts [like] bile [and] spermatic [ducts]; excrement is turned into hair, feathers, squames. Whether by attraction, expulsion &c. The author of all things in the body replies in one word, Native Heat; in all it is also present in all excrement; since nature directs native heat so that it concocts, attracts, expels.

Wherefore just as aliment for nourishing the parts, it is transmitted partly by the digestive faculty, partly it is attracted, partly it seeks of its own accord.

Likewise excrement partly seeks exit, partly sinks downward where an exit is provided; partly it is attracted, partly expelled; but preternatural excrement is not directed by nature, but with morbid malice it seeks different parts as bones, legs &c., lungs; so that as in the world so in the microcosm all things are moved to a particular place by their own will; wherefore those who say they are driven or are attracted suppose they are moved by another will. [According to] Aristotle, De Generatione, IV, 2, excrement

---

primarily to hold and stabilize the veins and secondarily to evacuate urine. Hence Harvey's phrase "fruit of trees."

[410] Du Laurens, pp. 254 ff.

[411] Duretus, *Magni Hippocratis Coaca Praesagia brevi enarratione illustrata decerpta* (Paris, 1657), p. 598: "It ought to be noted that it is generally assumed that there are three materials of urine, that is, drink and the more liquid portion of food, and humors of the liquefying parts, and properly speaking according to Galen after Hippocrates, urine is the serum of those humors which are in the veins, which have slipped down into the bladder and have been collected there, and this is the common material of urine and sweat."

according to its kind is excreted. Nevertheless there is motion in the body through attraction.

Preternatural For example: heat, pain, vacuum [as into] a cauldron; purger, chewer. Likewise motion is the same in that which extends downward; for example, tumors of the legs when they hang down and become very large. Likewise intertrigo always weeps; and ulcers of the mouth [result in] ptyalism[412] [which] in iij days can produce more humidity than [is ever contained in] the cranium. Δ also p.

Thus expulsion of urine occurs in all ways, [sometimes] 44v in none, partly in these, partly in those; [in] a preternatural [way] by attraction it appears in ulcers of the penis and of the bladder; [an irritation] as if a louse or ant had been introduced. By expulsion appearing in crises. Ischuria, dilation of the ureters and of the bladder. By downward extension, by liquefaction of the body where nothing irritates; nor is it attracted, for the force of the kidneys as of the prostate [is] one of expulsion.

[If] the body [is] constricted by cold [with] bare feet, [there is] copious urine. Likewise it appears spontaneously in the embryo where excrements spontaneously seek the cavity of the intestines; for [there is] no common function in the embryo, but according to its nature from native heat which is more powerful in retaining than attracting. There will be bile in the bladder after all [other] things have been purged.

During birth the humor is diverted for relaxing the parts; through vomiting, urine ΔΔ; coagulated; *Blood in ye fawces. Man in gibbets.* Δ in ulcers, Δ in dropsy, Δ in sweats, Δ in sweaty hands; ptyalismus Δ by acidulous waters, by juice, flax seed in the mouth, by fracture of a joint. Δ by walking about on a cold morning; mercury having been taken; urines and dry wine; Δ by timidity, lust *piss often hors befor ye rase.*

The number [is] two because bipartite, right and left, 45r because of magnitude of their work and for the sake of security; wherefore one having been removed the other

---

[412] A profuse discharge of saliva.

118

suffices.[413] Rarely iij or 4 [3 according to] Eustachius and [according to] Du Laurens 4. I saw only one, Piccolomini [described] one;[414] one [according to] Bauhin in a middle site.[415] Fat [according to] Aristotle.[416] The remainder when filtered is serous, for at the end of sanguineous coction [only] fat [remains].

Two membranes, one looser from the peritoneum where fat. The other special without fat, much thinner.

Shape [the same] as in sheep, bean-shaped, *kidney beenes*; internally partly gibbous, partly concave where the ij said sinuses for the hollow where the emulgents enter the prominences; in the middle it makes 2nd. In some there are no sinuses, but the kidneys divided into lobes as in young bulls and calves; in bears; and see the embryo; twice,[417] 1st *more*, 2nd *less*; Eustachius saw in the seal a tesselated placenta [=like structure] which Belon[418] [mistook] for the

[413] This interesting observation suggests that Harvey may have been thinking of the desirability of nephrectomy—impossible, of course, in his day. The priority for deliberate nephrectomy is usually given to Gustav Simon (1824–1876), who described a case in 1870, although an earlier case had been described by Erastus Bradley Wolcott (1804–1880) in 1861.

[414] Eustachius, *Opuscula Anatomica* (Venice, 1564), p. 51; Du Laurens, p. 252: "sometimes I have observed three and four." Supernumerary kidneys are rare, and Du Laurens was almost certainly in error in describing four. A single kidney is not so uncommon. Modern figures give an estimate of about 1 in 1,200 persons. Harvey says he saw such a case, as did Piccolomini, *Anatomiae Prelectiones* (Rome, 1586), p. 140: "I saw . . . a single, notably large one."

[415] Bauhin, pp. 152–153: "In a public anatomy which we conducted in the year 1589 we discovered a left kidney of remarkable shape and site, as well as the emulgent and spermatic vessels. . . . The kidney was above the division of the aorta and vena cava near the sacrum in that cavity in which the bladder was located. . . . In that place where the left kidney ought to have been located, nature had substituted a glandular and fatty substance." This refers to a horseshoe kidney, the original account of which was given by Berengario da Carpi (*Isagogae Breves* [Bologna, 1522], f. 17ᵛ), who had seen one in 1521 that was "continuous as if one kidney." The anomaly was also seen and noted by Andres de Laguna, *Anatomica Methodus* (Paris, 1535), f. 28ʳ. Thereafter it was seen and described in detail by Leonardo Botallo in 1564 (Benjamin and Schullian, *J. Hist. Med.* [1950], 5:315–336).

[416] *P.A.*, 672a:1: "Of all the viscera the kidneys are those that have the most fat. This is in the first place the result of necessity because the kidneys are the parts through which the residual matter percolates. For the blood which is left behind after this excretion, being of pure quality, is of easy concoction, and the final result of thorough blood concoction is lard and suet."

[417] In the human foetus the kidney is lobulated as it is in many animals.

[418] The published transcription erroneously gives Bellini, who, however, was born in 1643. Rather, Harvey wrote Bell[onius], that is, Pierre Belon

119

spleen X. Sign that the ureters from it; in another female not tesselated.

Size: length of 4 vertebrae, width of iij fingers. Sometimes they are not equal in size. In some larger, in others smaller. Quantity of fat less on the right than on the left, because the right part drier.[419]

Use [of fat] to enclose the sharpness [of the] *urine*, to aid coction, by softness and blandness; to conserve the kidneys, by warming the kidneys.[420]

WH I believe it heats the kidneys and *chookes them*; wherefore flatus is detained [and] excrements are suppressed; wherefore in the obese and nephritic more pains, especially in a hot constitution.

Veins and arteries; it milks out two adipose veins.[421] Right adipose rarely from the trunk but from emulgent; left, on the contrary, arises from the [vena] cava in the exterior tunic. Irrigating by numerous branches. Branch to the gland of Eustachius.[422]

A branch, sometimes ij, is inserted into the emulgent 45v from the azygous [vein],[423] wherefore material from the thorax [can pass] into the urine. Very obscure nerves from the 6th pair descending through the back,[424] inserted into

---

(*ca.* 1517–1564), and was apparently thinking of Belon's *Histoire naturelle des estranges poissons marins* (1551), in which, according to F. J. Cole, *History of Comparative Anatomy*, p. 61, Belon "saw a divided spleen but could not find a caecum."

[419] *P.A.*, 672a:23: " . . . the right kidney is less fat than its fellow. The reason for this is, that the parts on the right side are naturally more solid and more suited for motion than those on the left. But motion is antagonistic to fat, for it tends to melt it."

[420] *P.A.*, 672a:16: " . . . the fat has a final cause, namely to ensure the safety of the kidneys, and to maintain their natural heat." *Ibid.*, 672a:20: "The kidneys, moreover, by being fat are the better enabled to secrete and concoct their fluids; for fat is hot, and it is heat that effects concoction."

[421] The perirenal veins.

[422] The suprarenal glands. These glands were not seen or illustrated by Vesalius but are shown in detail in the plates of Eustachius.

[423] There is no normal communication, as such, between the renal and azygous veins. Probably here Harvey is confused by the lumbar azygous veins which, when present, are situated close to but behind the renal veins. Harvey's statement that "material from the thorax [can pass into] the urine" is manifestly wrong, for the direction of the flow of blood in the azygous veins is *upward* toward the heart.

[424] Here Harvey is referring to branches of the vagus nerve.

120

the special tunic, wherefore [there is] a relationship [with] the stomach in nephritic vomiting, because the same branch ascends to the mouth of the stomach.

Nerve also entering with the emulgent artery from the plexus of nerves,[425] from beginning of the mesentery; nevertheless Fernel [states] and experience testifies that the kidneys have little sensitivity,[426] wherefore impostema,[427] and they are corroded and consumed with either none or slight pains, rather by affection of the adjacent parts than their own.

The gland of Eustachius [suprarenal gland] in the superior part where it looks to the vein adheres firmly to the anterior membrane.

[This is regarded as] the receptable of melancholy by some. A gland of this sort Δ WH in some embryos much larger than the kidneys themselves.[428] It may not be present in the dog.

Site

In embryos [it is] large, lax and filled with blood; a spongy cavity like a bladder of blood or a sponge filled with blood. It seems *the nerer the birth the bigger*. Wherefore per l'envoy I formerly thought it [to be] a pustular excrement and an adipose vein to carry blood thither; also I have thought [it to be] from the [vena] cava.

Shape

Ginney cunneys in another color, *straw color. In a ratt milke-white*; in man very obscure [color]. In man either it is obliterated or worthy of note. NB Often when the kidneys have been removed it adheres to the diaphragm;[429] wherefore use is only seen in the embryo.

Substance

---

[425] This refers to the renal sympathetic plexus, particularly the branches from the coeliac plexus and ganglion coming, as Harvey says, "from the beginning of the mesentery."

[426] Fernel, *De Partium Corporis Humani Descriptione*, in *Universa Medicina*, (Paris, 1567), p. 19.

[427] Apparently a hybrid formed from the English word "impostume" or "imposthume," common in Harvey's day, and the Latin *apostema* (abscess).

[428] As Harvey points out, in the foetus the suprarenal glands may appear relatively larger than the kidney.

[429] This is an important point, for when the renal capsule is incised the kidney can be removed without the suprarenal being seen, for the suprarenal is embedded in the fat above the kidney. This may well explain why the gland was not seen by the earlier anatomists, including Vesalius.

The right larger than the left. Sometimes it imitates the shape of the kidneys, sometimes not.

[There is] a divided insertion of the emulgents [with] 46r two branches on each side into two sinuses and thence divided. Four individually into four regions of the kidneys; of arteries as well as of veins; they are stretched out almost as far as the gibbous part.

[The kidney has] a hard, compact substance very much like that of the heart; of obscure reddish color; temperament hot and moist, inasmuch as sanguineous flesh.

Insertion of the ureters in the lower sinus,[430] whence divided into many wide branches which *grow larger and larger*. They receive the papillae in wide orifices. They void urine as infants milk; wherefore expulsion of urine rather by attraction of these than of the kidneys; wherefore kidneys, just as breasts for the infant, serve to receive the moisture, which must be sucked out through the papillae.

Number of the papillae uncertain: 9, 10. They receive the papillae at the top of the sides rather than in the end; wherefore in the sheep all the papillae [arise] from a common sinus; whiter color of the papillae, *soaked with vrine*; wherefore *the* inner portion of the kidneys *wors tasted then without*.

There are passages in the papillae for the urine but so narrow that [when] collapsed they are invisible and do not admit a hair. Here in the ureters[431] occur the calculi of the kidneys; and they fill the papillae wherefore a mixture of blood; ulcers; dregs of the vein; also through anastomosis; wherefore shape of calculi *as cast in this mowld*.[432]

Galen 4 [According to] Fernel all calculi here rudimentary;[433]
Aph[orism]
Hairs produced   [430] This refers to the renal pelvis and the calyces. Each renal papilla opens
from a dried-   into a minor calyx which join to form a major calyx. There may be up to
out nature[434]   thirteen minor calyces and usually three major ones.
     [431] Harvey uses the term ureter to mean the renal pelvis.
     [432] A so-called staghorn calculus takes on the precise shape of the renal pelvis and calyces. In Harvey's time such calculi would have been common.
     [433] *De Partium Morbis et Symptomatis*, in *Universa Medicina* (Paris, 1567), p. 309. Small calculi formed in the renal pelvis pass down the ureter to the bladder, where if not voided they become enlarged.
     [434] Galen, *Comm. IV in Hippocratis Aphorismos*, 76; Kühn, XVIIb, 769: "Bodies similar to hairs are generated by such a dried-out humor in the kidneys."

Kernell; increased in bladder. Here I have seen worms in the dog so that hawks always [attack] in the back.

Account: *here commeth owte of the veyne and artery the* 46v *serus substance excrement* of the second coction [which is] *proportionable to the sweat; by these Branches passed into ye kidney by the papillae* &c. &c.

Passions and diseases; kidneys flaccid and without blood *as parboyled* likewise *russetish* in cachetics and hydropics. WH I believe those who are cold and dry of intemperate constitution.

Aposteme of white, fetid matter; wife *Chirn* with ureter distended by pus;[435] like intestine corrupted by ulcer; like a purse according to Fernel[436] and Laelius a Fonte.[437]

Black gangrene. NB Intermittent fevers become inordinate by these apostemes toward evening, especially in those afflicted likewise with brief horripilations. *Sr Robert Wroth.*[438] Likewise fevers and a very great unevenness but without pain; but grave distention and very blunt pain[439] from tension of the external tunics.

Ureters [run] from the kidneys to the neck of the blad- 47r der wherefore in those [animals] which lack a bladder also ureters [are lacking]. WH They seem rather to arise from the bladder than from the kidneys, because they cannot be separated thence as easily as from the kidneys. Because of the membraneous substance of the bladder [they are] white, exsanguinate, sinewy. They seem more necessary than the bladder, and the bladder is for their sake [and acts

[435] Mrs. Chirn quite obviously suffered from a pyonephrosis probably following on renal calculi.

[436] *De Partium Morbis,* in *Universa Medicina* (Paris, 1567), p. 310.

[437] Not "Coelius" as in the published transcription, but Laelius a Fonte *Eugubinus,* fl. end of sixteenth and beginning of seventeenth centuries, practiced in Rome and Venice, and published his *Consultationes Medicinales* (Venice, 1608); however, Harvey employed the second edition (Frankfurt, 1609), cf. n. 511, and his reference here is to p. 241: "In a cadaver dissected by Angelo da San Lopido the left kidney was found to be like a purse without substance but only a covering filled with little stones and ichor-like sanies."

[438] Sir Robert Wroth (ca. 1576–1614), not Wrath as in the published transcription, was the son of an important parliamentarian, and husband of Lady Mary, niece of Sir Philip Sidney and author of *Urania,* whom he married at Penshurst in 1604. He was Sheriff of Essex in 1613.

[439] A good description of the dull pain produced by distention of the kidney.

as] a diverticulum of the urine until time for its voiding. WH Wherefore perhaps the ureters in birds. Wherefore it is possible to live for a time without a bladder, but not without ureters. Wherefore the insertion into the neck of the bladder [and] at the root of the penis. Veins, artery from neighboring parts, and nerve surround [the ureters]; thus [there is] keen sensitivity and pain; wherefore here occur tortures[440] when a calculus impinges on them. I have observed a double left [ureter].[441]

Number, shape, quantity. In women, wide, short, straight; the contrary in men: thickness of straw, length of iij palms. Thus a very small calculus can pass through [although] they can be much dilated as I have seen the intestines distended by pus. Here occurs obstruction with nephritic pains. Wherefore [according to] Plater rupture from a calculus and the urine flows into the bowel.[442]

Stone Wherefore [in the case of] Dr. Argent stone eat out his passadge in ye flanke;[443] but sign that neither acrid urine nor hot body from color of calculus; almost gypsum-like pituita; otherwise it would cause fevers and inflammation and death. The calculus sticking and the ureter distending in its upper part[444] with the lower part contracted, it becomes inflamed above the calculus.

Genitalia in general. by the string tyed to eternity; where- 47v fore since nature was not able [to produce] individual eternity, yet by the faculty of these parts it was able [to produce] a kind of eternity by generating [species] similar to themselves through the ages. Wherefore in the sacred scriptures greatest blessing Issue that thy seed shale remayne for ever. For it is seen as divine, eternal, perfect renewal; by lacking rather than by enjoying [individual eternity]; that each thing from which [there is] no progeny is de-

[440] Harvey would have seen many patients with renal and ureteric colic.
[441] Duplication of the ureter on one side occurs in about 3 per cent of individuals.
[442] Plater, p. 165 [157]. A fistulous opening between the ureter and the bowel is rare. The fistula is usually between the bladder and the colon or rectum.
[443] See note 291.
[444] Here Harvey is referring to hydro-ureter and hydronephrosis resulting from an impacted calculus.

124

liberately terminated here; on the other hand, verisimilitude to others to remain eternally. That men and dogs now live [descended] from the loins of those who [lived] a thousand millennia [ago] and ages hence.

Such the necessity that nature solicitous that the individual generate similar to self; for the sake of the species wherefore (*as Nature regarding* him not) immediately *despayreth declineth* and if these parts, indeed, *these little strings* were lacking to all men *or gelded*, it would affect the human race.

Hence *how greate and most affecting pleasure* nature [granted] for this action, wherefore men impetuously pursue, seek, and perform what per se *lothsome;* wherefore *as* nothing *more pleasing to them which desier or act*, so nothing *more lothsome to them which ar past it or come to see it.*

Hence nature [provided] a balance of the lustful with sterility and difficulty of generation and very often in vain; wherefore among them the greater lustfulness with which the act is very often performed [is] among animals and among men of a hot region. And so the action of those parts to generate similar to self, to plant man properly by sperm, to make, preserve and provide for.

Services: 1, perpetuity. 2, the mind is sharpened and the 48r body is rendered healthful; wherefore about the advent of sexual desire either they are cured of an incurable disease; or some are not; or they begin to be incurable; 3, the vigor of the body is aroused by the little flame of these parts; vigor appears to the castrati of whom the nature, spirit, temperament and mores are so altered that [they are] less than all in magnanimity, talent, wisdom, courage and health; they decline into women. Likewise it appears to males and females [formerly] moderately friendly, *never more brave sprightly blith valiant plesant or bewtifull* than now that coitus is about to be performed.

It appears from those things which occur around the onset of venery: the voice is changed, the groins become pubescent, the breasts full, [and] charm of countenance is united to comeliness in all the members.

As much as things taken in moderation invigorate, too

125

much enervates and the mind becomes dull, because *by how much this is affecting pleasant*, too much abused [is] very dangerous; as wine, in which great service for evacuation of excrement, by lack of which men suffer greatly; wherefore it is permitted to clerics for expurgating the kidneys. One suffering spasmodic distortion ought not wait; I have praised a French bawd. Likewise in women, much graver things from histories and those things newly expressed and daily [by those] who cure hysterical symptoms.

Sperm because [it is] the particular effective of generation; all these parts for the sake of the sperm; and as many as they are [in men], more in women; wherefore the perfection of this operation, viz., generation, about which nature is so solicitous, that in the whole anatomy nothing more admirable than the structure of those things. The more they are requred for perfection the more divine, and on the contrary the more excellent the animal the more perfected. Elsewhere I have written of the female parts where in particular the anatomy of the pregnant.

The genital parts about the sperm. Others: preparantia,[445] ij veins, ij arteries, [in] n[umber] 4; they elaborate, prepare, make fecund. Testes, deferentia fecund from testes, the reservoir and vesicles of the prostate[446] preserve, the glans [serves for] exciting lust, the penis [for] ejaculating. 48v

Some of these parts absent in some, all most perfect [in] *ratts* wherefore potent.

Degrees of generation of animals

1. some from putridity by heat of sun and moisture of earth.[447]

2. some from sperm as plants from seeds, *oysters, musles, flyes* &c.; wherefore here neither male nor female

---

[445] Cf. n. 135. *Vasa preparantia* was the old name for the spermatic cord or more particularly for the vessels in the cord. The term deferentia was used for the epididymis.

[446] The vesiculae seminales are, of course, separate from the prostate, but Harvey seems to regard them as part of the prostate.

[447] Here Harvey is postulating spontaneous generation, which at this time was the current teaching. Later, in *De Generatione* (1651), he was to deny the possibility of spontaneous generation.

but proportionate, as *Thistle and seed hemp*, which do not copulate, but spermaticize; *spatt*, in place of the female, in which [they spermaticize] outwardly [and use] the earth or water as the womb. But from these things it may seem as if coitus [occurs]; they do not generate or at least another thing which, as in the case of the nits of the louse, either not animal or different from things which generate; as in the case of the *Caterpillar* without others; caterpillars [or] small flies through continual metamorphosis.

3, others male and female, which in the former or out of the latter; the male in which the formative force [is] the active principle;[448] the female in which [is provided] the place and material; thus the origin of generation begins as male and it is completed as female. Male *woe allure make love*; female *yeald condescend suffer*; the contrary *preposterous*. The fabricator seeks material; like heat, celestial, infernal.

Regarding sperm what it is; how it is made; and whence; what it contributes to generation and in what way productive, and of the genital parts.

I have transcribed

The spermatic vessels [have their] origin [from the great] vein and artery, 2, because they arise from them and carry blood; 3, and send forth branches to the neighboring parts.

49r

In general

4, and in the embryo they are very broad as far as the testicles.

Wherefore in some [there are] 7 vessels; of the vessels

5, texture, function and contents; wherefore sperm not from the whole body but [only] from these parts.

[There is] division by which the sperm [is] more perfected, longer because of delay and thence [better] coction and reduplicated; for which the preparantia and deferentia[449] [are required].

In the posterior part of the bladder X with the ureters for joined together in the glands

In which, on the other hand, only the deferentia, as in fish,[450] from the diaphragm to the anus.

[448] Harvey is putting forward the then generally accepted theory of Aristotle that the male provides the semen but the female provides nothing to the foetus except the place where it is to grow; the foetus being developed from a mixture of semen and menstrual blood.

[449] Cf. n. 135, f. 48ᵛ, and n. 445.

[450] *G.A.*, 765:34: " . . . many animals in which the distinction of sex exists, and which produce both male and female offspring, nevertheless have

127

1, in the middle of them, the testicles and parastati[451] reduplicated. The liver and heart and the testicles joined to them, and in the same way the liver and testicles.

2, wherefore in women uteri and urine extend to the testicles. In some females to the diaphragm; wherefore in birds the first conception [is near] the diaphragm; afterward when completed, descent [occurs].

Wherefore properly the spermatic vessels are the deferentia[452] and the special constitution of them, and contain sperm.

2, **WH** nevertheless in *ratts* I have seen blood contained[453] and perhaps sometimes so in man; wherefore it is said that blood is emitted in coitus; for crude sperm is blood; as menstrua easily by touch from the preparantia.

3, as *a lutestring hard rownd slipping from your fingers;* wherefore varicose hernias are diagnosed and inflammation of the testicles *cowtts evell,* for distended by putrefaction of sperm in lues.

4, by flatus which is a sign of sterility for Aristotle;[454] the groin swollen by sperm, *cowttes evell.*[455]

5, they and the testes are retracted filled with sperm; sign of health. Because perfected by coction from which [comes] sperm.

Preparantia, veins and arteries, joined at the testicles. 49v

no testes. . . . I mean the classes of fish and serpents." Fishes, of course, do possess testes, but Aristotle preferred to regard the structures as spermatic ducts. Aristotle denied that the testis produced the sperm. To him it was merely a collection of tubes for conveying sperm already made in the vasa preparantia or spermatic vessels. It then passed from the testis into the epididymis (vasa deferentia) and then to the vesiculae seminales.

[451] Cf. n. 225. The ancient term for epididymis.

[452] Here Harvey uses the term deferentia (epididymis) instead of preparantia for the spermatic vessels.

[453] It would seem that Harvey here wonders if the spermatic vessels do contain sperm in accordance with the old teaching. He is cautious not to say that this always occurs in man, merely contenting himself with the statement "perhaps sometimes in man." However, he goes on to say dogmatically that sperm is developed from blood.

[454] *G.A.,* 746b:25 ff.

[455] I.e., colts evil, a term in veterinary medicine; Italian farriers in the seventeenth century used the term for priapism, but English farriers disagreed and applied it to any swelling of the penis, sheath, or testicles. See Gervase Markham, *Markhams Masterpiece* (London, 1610), bk. I, ch. 87: "Of the colt evill."

128

The right vein a little below the origin of the emulgent from the superior part of the vena cava and from a long thickish root [placed] anteriorly. Galen notes a branch to this from the emulgent. The left vein from the lower part of the emulgent to avoid the motion of the aorta. Hence it was said sperm from the right [produces] males;[456] the contrary, females, because more concocted from blood of the [right] vein, and it generates more perfectly a more perfect being; viz., the genitalia of the male.

WH The same but not for the same reason. Because the artery [comes] from the trunk of the artery; because not the more crude blood of the emulgent; perhaps [related to] the side, because the right more perfect than the left and the blood of the right in the same way; of this later.

Site. Each artery from the large trunk of an artery, very far from the emulgent; a little above the mesenteric artery which, like the others, [is filled with] spirituous blood and is joined with the vein by fibrous connections behind the right *ridg over the veyne*.[457]

and soe both joined obliquely *on this Ridg* at the produc-   Site
tion of the peritoneum where with a nervule of the sixth pair and the cremaster [muscle], they go to the testicles;[458] whence reflected by the same course, it forms the deferens and here again entrance into the neck of the bladder to the root of the penis; wherefore both [employ] the same passage, [but] they differ in position *figure* and substance, wherefore different names: *streetes* as *I sayd in ye gutts. from smith feild* &c.[459] Riolan 349[460] from Galen d[e] u[su]

---

[456] Cf. f. 23[r] and n. 218.

[457] The right testicular artery arises from the aorta usually at about the same level as the inferior mesenteric artery, but the left one arises above this artery.

[458] Harvey apparently uses this term to mean the processus vaginalis testis. It is used in this sense too, by Culpeper in his translation of Riolan's *A Sure Guide* (London, 1671). Motherby uses the term *productio* as synonymous with an apophysis or appendix.

[459] Cf. f. 20[v] and n. 192 for further references to London's geography.

[460] Riolan, p. 349: "For if you blow into one spermatic vein you will observe the other inflated with the arteries, which was first noted by Galen." This reference by Harvey to the page of Riolan's work permits us to determine that he employed the edition published in 1626, and hence indicates that he was at work on his notes at least as late as this time.

p[artium]: 16 ch. 12 by insufflation the spermatic veins are inflated with the arteries; WH or the contrary?

Worthiness The testes which attest virility, worthy to be eaten, are 5or called principal bodies, the principal ones of the genital parts, because they make the productive semen. WH *a little staggerum* [i.e., hesitancy] *in these.* 3 [considerations]: 1st, without them the body can live; those castrated decline into female; not by lack of these but of sexual desire; *Jentel pleasing divine heate* worshipped for the goddess Venus. Therefore with the whole body these parts are lacking in vigor, which occurs to men in old age; [but] to birds in the spring; and by the supply of nourishment; wherefore these parts attest the advent of sexual desire when they are augmented and pubescent. Wherefore lacking in birds except at the time of coitus. But sexual desire without use weakens as in those in whom, by abstinence and by chastity the testicles are diminished, and the penis is retracted into the abdomen completely cold; in the stag it putrefies; in the hare the testicles do not appear.

2nd, in fish and serpents there are none at all,[461] wherefore it appears that they are not necessary like the liver and heart; for having been cut away [the animals] live, but not simply and for the sake of the species. Although it may be that by castration the body loses vigor, it is not because vigor is in the testicles, but because by lack of the instruments the faculty weakens; the veins having been retracted and not extended, they are obliterated; Δ umbilicals; and because redundant that superfluity is turned into fat and its nature altered; it happens to them Δ *spring or sumer overclowded nothing ripens.*

From the beginning of puberty to old age [is] the age of life; wherefore the castrati [are] like that region in which [there is] no summer, only spring, autumn and winter. Therefore *Eunuchs* always appear either like boys or little old women without youthfulness; from lack of concocting vigor nature turns them from childhood into the character of women.

[461] *P.A.*, 697a:9: "No fish has testicles either externally or internally." *G.A.*, 717a:19: "neither serpents have testes nor have fish."

The principal parts of generation [are] like the *Cristall*[462]
in the eye; they are not in all. Δ *analogous.*

2nd [i.e., 3rd], like appendices to necessary parts, so that [according to] Aristotle, [although] attached they hang in that way in which the weights of cloth looms are attached; these having been removed, the passages are drawn inward, so that (he says) the testes make a more stable movement of the [spermatic] secretion.[463] Therefore in all [there is] a vas deferens in which the sperm [moves]; in certain ones it is the principal part of the testicles and is always present, viz., the vas deferens. The remaining [parts are] more posterior, like the vesicles, mentula,[464] prostate; so the testicles, viz., for well-being or necessity, [are like] the eyelids on the eyes, not as the *Cristal.*

Δ As the weights of looms [move] *evenly to stretch lest long passages intertangle.* Δ *making Ropes twining of silk, making bone lace without that they would run together and intertangle.*

So the testicles hang anteriorly, not entirely *fixd* as bones but *sufficient to stretch forth & yett yeald;* because the swelling passages are retracted; Δ *rope maker moving as the Rope twines.* Therefore the testicles, now pendant, now retracted, inward, outward. Thus in them the vessels of greater width; there the testicles less fixed; Δ in man; *Hares ratts,* where on the other hand, in a smaller [animal] attached to the nates; *Boare,* inward as in birds; in women.

In those having testicles they are attached to the middle of the vessels. Wherefore in those in which testicles [are present, there are] preparantia;[465] those in which [they are] not [present have] only deferentia; Δ fish, serpents;[466] wherefore in those in which [there are] no testicles, the

[462] The crystalline humor or lens.
[463] *G.A.,* 717a:34: "For the testes are no part of the [spermatic] ducts but are only attached to them, as women fasten stones to the loom when weaving; if they are removed the ducts are drawn up internally." *G. A.,* 717a:30: "[The testes] make the movement of the spermatic secretion steadier, preserving the folding back of the passages."
[464] A name sometimes used for the penis.
[465] Cf. f. 16ᵛ, n. 135, f. 48ᵛ, and n. 445.
[466] Harvey followed Aristotle in his belief that fishes and serpents had no testes; cf. f. 49ʳ and n. 450.

liver [and] region of heart in place of testicles; wherefore *from the* testicles to the liver, preparantia; from the testicles to the bladder, deferentia; accordingly the spermatic vessels longer (and that for the perfection of the sperm).

Testes either outward or inward, and in certain ones now outward now inward; Δ *rattes Hare*. Inward at kidneys, Δ birds Δ in embryo of stag; at loins *ginney cony grampos and a porpos;* they hang loose within the abdomen. In colder animals and weather, inward as opposed to outward; in man outward; in woman the contrary; wherefore [according to] Bauhin from man can be made a woman with the vessels and mentula retracted [from their] position and the constitution of the parts [altered].[467] On the other hand they appear in the *Elephant* and *porpos Hedghog.*[468] hot creatures have inward. WH *hott is to be vnderstood hot* in sexual desire. WH In some outward in others inward for each reason; and when inward the buttocks arouse sexual desire; and because of the length of the vessels [perfection of the sperm is greater]. That inward or outward because of the skin present on the scrotum. X Δ *ratts* now inward now outward; *Hedghog mowle all that want feete want stones;*[469] those [with testicles] outward engage in venery more slowly; on the other hand inward in birds, fish; [those animals] are more lustful in which inward; birds, fish.

The more temperate animals [are] those in which [the 51v testicles are] outward, for by more coitus they are loosened because more perfectly in them; a sign of sperm, length of spermatic vessels; for the sperm is more perfected from

---

[467] Bauhin, p. 181: "[The testes] are larger and hotter in man than in women, not by reason of site, but by reason of the temperament of the whole body which is colder in woman and hotter in man; therefore in men the more abundant heat forces them outward; but in women because of less heat they remain inward. This is true as is clear from accounts of those in whom the heat having been rendered more vigorous, thrusts the testes outward, and so it is said that a woman has been changed into a man."

[468] *G. A.*, 717b:27: "All the vivipara have their testes in front, internally or externally except the hedgehog; he alone has them near the loin." *H.A.*, 500b:8: "the testicles [of the elephant] are not visible, but are concealed inside in the vicinity of the kidneys."

[469] *G.A.*, 765a:34, says that the footless animals have no testes.

132

much blood, much elaboration. Δ *Starch* from[470] much flour [and] 1 ounce of oil; for so many years. According to all these things the opinion of Aristotle that they are like weights[471] [is] not so ridiculous.

But for another reason the testicles are of benefit and concoct and make productive; Δ in birds *partridg Quailes fessants melers*. *Hares* in which at the time of coitus the testicles are filled with sperm; just as in those in which no testicles the passages turgid with sperm; wherefore also *lambs stones sweete*, on the other hand, *Rames;* wherefore Δ in man by reason of ulcer the testicles without sperm. To the 3rd

to little eggs tyed to the vessells wherefore a connection with the veins so that application of cold to the testes cools them greatly; stops hemorrhage[472] and causes rigors. So the testicles having been inflamed, a cough, and alteration of the voice from augmentation of the testicles. Also a connection of a similar kind occurs with the brain. 52r
Testicles,
shape and
connection

Size of the testicles is not related to greater capacity for generation or to the amplitude of the virile member or vulva; for males given to venery and potent perhaps have greater quantity of sperm.

A certain motion of constriction and dilatation appears when anointed before fire, which I believe is [similar to] the motion of the uterus. On the other hand, flaccid in hysterics and in bodies with enfeebled intestines, *as knitting of weemens brest*.

In number, twin; [according to] Nicolò Massa one with very wide vessel on the right side, the left [had] neither vessel nor testicle.[473] Holler writes of an acquaintance in

---

[470] Here the published transcript erroneously gives "Starcheus," which appears correctly to be "*Starch* ex."

[471] Cf. f. 50ᵛ.

[472] The published transcription incorrectly gives *haemortrogiam sentit*, which correctly should be *haemorrhagiam sistit*.

[473] Massa, f. 36ᵛ: "They are two in number in all. . . . In one person dissected by me, I saw only one testicle on the right, and to it descended a remarkably wide spermatic vessel, but on the left there was no testis nor did any vessel descend to this place from the emulgent since it was not necessary, because nature had created him without a [left] testicle."

whom iij,[474] WH but fleshiness deceived him as to the third, or a tumor of the parastatae.[475]

[There is] a division in the containing tunics; cremaster muscles, nerves, vasa deferentia and preparantia, epididymi and their substances. Here no fat unless in the delicate: *lambs and cocks;* otherwise *make ill Juse.*

Diversity of authors regarding number of tunics, some [say] 3, others 4, 5, 6.[476] WH Fissile membranes and two made from one; wherefore [according to Du] Laurens all membranes double.[477]

Anatomist too curious; making mor parts then is to noe end. The number truly four, clearly two common, two special. 1st scrotum, skin, purse, *lik a lether purs with too pens;* soft, wrinkled so it can easily be distended and contracted; separated by a seam.[478]

NB In paracentesis avoid the raphe because of bad re- 52v sults; like tendons of muscles it is consolidated with great difficulty [if cut]; great pain and gangrene; recently happened in a man.[479]

Scrotum is made of skin and epidermis; thinner, laxer and softer, without fat, than in other places; but so that where it is separated from the underlying flesh *as in Elboe,* wrinkled.[480]

Dartos, *the second coate,* is easily separated from the other; fleshy portion of panniculus, fleshy fibers without fat.

EluTROIDES vaginalis, first of the special [tunics], *like*

[474] Jacobus Hollerius, *De Morborum Curatione,* (Paris, 1565), p. 173: "In December 1564 while I am writing these things in Paris, I chanced to see a boy, son of a printer, who, as all believed, had three testicles." For another case, cf. L. Belloni, "Ueber das Wappen und die vermeintliche Triorchidie von Bartolomeo Colleoni," *Centaurus* (1950), 1:43–61.

[475] Harvey uses parastata here to mean the epididymis.

[476] If everything from the skin inward is counted, there are, in fact, eight layers covering the testis. Harvey restricts it to four: skin, dartos muscle, tunica vaginalis testis, and tunica albuginea.

[477] Du Laurens, p. 267.

[478] Refers to the median raphe of the scrotum.

[479] The published transcription erroneously gives *Nov: Jo: Seiton,* which should be *Nov[iter] ho[mini] fectum.*

[480] The published transcription erroneously gives *sep,* which correctly should be *rug[osa].*

134

a long scabberd, [is] a process of peritoneum;[481] here it is attached point d'ore; by relaxation of this hernia [occurs]; wherefore to cure, compress this, or consolidate or bind. Exteriorly it is connected to the dartos by fleshy fibers; inwardly it is covered by a watery humor and by veins. Hence it is counted by certain ones for two tunics, because inwardly and outwardly [it has] a different constitution. Its fibers [are] like fleshy muscles [and] retract the testes as in dogs. The use of this tunic [is] to connect the vessels to the testes and to each other, in which **WH** sometimes so much fat that I have been unable to separate the vessels.

The albuginea, a special [tunic] peculiar to the testicles. It arises from a tunic of the spermatic vessels. White, thick, strong, to strengthen and bind the soft substance of the testicles.

A controversy regarding its name, for some **X** call it epididymis.

Cremaster muscles [or] suspensorii on each side longish 53r and rounded; arise from a strong ligament of the bone of the ilium which runs from the spine of the ilium to the pubic bone where the transverse muscles are terminated; they appear as parts of these and they enter through the tendon of the oblique foramina.[482]

Extrinsic near the inguinal region, they are attached to the spermatic vessels at the head of the testicles.[483] The elutroidae tunics of Vesalius.[484] Sometimes from the interior of the pubic bone fleshy fibers are communicated to them, wherefore a twin origin (as in apes) is seen.

A nerve from the 6th pair, which [is] near the root of the

[481] The processus vaginalis testis is, in the foetus and baby, a tube of peritoneum. After the descent of the testis, the cranial portion of the tube obliterates, leaving the tunica vaginalis surrounding the testis. Should the processus remain patent, congenital inguinal hernia can occur. It would seem that Harvey was quite familiar with this condition and its cause.

[482] The cremaster muscles arise from the fibers of the internal oblique muscles which arch over the spermatic cord. As Harvey says, this comes from the inguinal ligament ("strong ligament of the bone of the ilium"). The oblique foramina refers to the inguinal canal.

[483] In fact, the cremaster muscle and fascia form a covering for the testis.

[484] *Fabrica*, p. 448.

ribs [supplies them];[485] *of all creatures Ratts have these partes most playnely delineated and curiously made* whence I believe them so generatively potent; wherefore [rats are] not arisen from putridity as commonly thought. Because nature does nothing in vain, especially so elegantly.

The vasa preparantia are inserted in the testicle where the artery and vein form one body; pyramidal [in shape]; from innumerable little branches of arteries and veins joined by anastomosis, by which it is interwoven and a plexus is formed. A varicose body, vine-shaped in form. Here the color alters, but the materials are transmuted; *In a ratt like Infinit little gutts.* From here they descend through the testes, not **WH** however in birds at the time they engage in coitus.

Thence again upward, twisted about they resemble the shape of a goat's horn, wherefore parastati [according to] Herophilus.[486] From here the vasa deferentia[487] take origin.

Parastati [or] *as Fallopius* [calls them] epididymi.[488] The 53v testicles connected only at the head and base; no transit; nevertheless in some the sperm [passes] into the testicles. The middle part of the spermatic vessels, which [dissected out] extend to the knee, white, long, thick, by degrees narrower; of which the origin in the vas deferens; [cf.] Vesalius, Plater;[489] *figure in a Ratt;* [they are twisted] as the intestines [are] convoluted. Substance between very soft vessels and hard testes.

[485] The sixth pair of nerves referred, in Harvey's time and before, to the modern glossopharyngeal, vagus, and accessory nerves. The nerve in the inguinal region to which he refers here is the ilioinguinal nerve from the lumbar plexus and is not a branch of the "6th pair," which, of course, are cranial nerves.

[486] Galen, *De Utilitate Respirationis*, ch. 15; Kühn, IV, 565.

[487] The term vasa deferentia is used here in its modern sense.

[488] Fallopius, f. 187ʳ: " . . . by the word EPIDIDYMIS we ought to understand something which, although small, is similar in substance, shape and use to the testis and finally, as the prefix EPI requires, rests upon it. But, you enquire, what is this testicle? I reply that it is the longish body attached to each pole of the testis in its indented part, that is to say, at its head and its base." Folio 187ᵛ: " . . . the same substance and use as the testis— for Galen agrees that semen is made in it and that it is often found filled with semen—and does not differ in shape; we can properly call it a testicle, that is to say, DIDYMIS: and since it is attached to the testis itself, we shall also call it the EPIDIDYMIS."

[489] *Fabrica*, p. 450; Plater, p. 159.

136

Passions: filled with sperm wherefore the swellings of the testicles; *the cowttes evell*, and from gonorrhea such a huge mass that goes about the testicles. Which is apparent because the testicles hang down evenly and are not so separate to touch.

Tumor commonly in inferior part toward the [vas] deferens because thence virulence and there a hard ganglion will remain[490] after the tumors cured; wherefore sometimes when they are turgid with sperm [there appears to be] a third testicle, as I have seen. [According to] Nicolò Massa larger than a testicle.[491]

Observation. Here in addition to rupture four things are common: varicose hernia when the vessels have been distended as *Chickens gutts descends*, which sort I have cured although the cure is very difficult. Likewise fleshy hernia from sarcosis, *a fungous flesh* or parenchyma, commonly *Carnosity*.

Sometimes [they are swollen] with a very great quantity of water and flatus; *the man behind covent garden bigger than his belly;*[492] *forme*, penis as if of a buffalo.

Common ruptures which we have seen cured very often: puerocele in boys; I have easily cured bubonecele; *the woman baggs*. Hydrocele: especially dropsy I have often cured. Epiplocele and eterocele[493] I cured in an adult man.

Substance or flesh of the testicles, white and soft; *pappy* 54r *wheyish, glandulus curdy*, a little rarer, laxer and thicker than the cerebellum. *Ratts like a bottom of threed plitted together* in infinite networks.

Diverse opinions of authors regarding temperament. WH Cold because exsanguinate; substance glandular, milky, very similar to cerebellum; as if made from the same material [according to] Galen;[494] observation: how do they concoct? reply that [like] the milk of the breast by

---

[490] A residual swelling in the epididymis after epididymo-orchitis is a common finding.

[491] Massa, f. 36ᵛ.

[492] An interesting case of what must have been an enormous hernia plus hydrocele.

[493] More correctly enterocele, a hernia containing bowel, in contradistinction to epiplocele, which is one containing omentum.

[494] *De Usu Partium*, VIII, xiv; Kühn, III, 678; Daremberg, I, 565.

inflowing heat; humid also because soft and lax, wherefore they readily accept spermatic humidity and retaining [it] are humectified.

The right hotter than the left, wherefore males [are generated]; but X Aristotle, males created with the right tied off.[495] WH Male from vigor of spirit *higher sett*, wherefore *bastards brave men* because created by greater heat in forbidden coitus; wherefore Diogenes, a drunken father procreated you; thus they withdraw stallions so that inflamed with greater desire they may impregnate better; wherefore not only these things but [according to] Aristotle the constitution of the air [is effective]; wherefore [according to] Aristotle greater internal dryness to procreate females;[496] wherefore [one should consult] a horoscope if proportional parturition [desired]. [In addition] to these things, hot testicles, size, pubic and belly hair and large veins in the scrotum [indicate] the more lustful and potent; wherefore those of hirsute belly [are] lustful like birds; on the other hand, those more humid by a humid seminal coction [are] productive; the drier [ones are] *Barren*.

Commonly; Δ in a virgin a double bladder.[497] WH Lord Chichester[498]

The urinary bladder *blather; blown vp* OURIDAKOS, 54v little daughter of the body, wherefore *urinal* same shape. In man who drinks more in proportionately larger quantity, for as I said the parts of the stomach are dilated by use. For by more copious [drinking] it can be distended; *(as children doe) by rowling and breaking the fibres*.

Shape, use, after opening. Concave [and] *narrower and*

---

[495] It would seem that Harvey has his doubts of the old teaching that the male foetus was developed from sperm coming from the right testis; cf. n. 218.

[496] *G.A.*, 766b:34: "more males are born if copulation takes place when north than when south winds are blowing," since the air is more moist when the wind is in the south; "the animals produce more secretion, and too much secretion is hard to concoct; hence the semen of the males is more liquid."

[497] Riolan, p. 247: "In some a double bladder was found, or at least a bladder divided into two cavities by a membranous septum. Volcher Coiter found a double bladder in a virgin 35 years old." This would be a case of diverticulum of the bladder.

[498] Arthur, Baron Chichester of Belfast, Lord Deputy of Ireland. The note must have been written after his death in Ireland on February 19, 1625.

narrower to the neck. It receives urine, retains and ejects at will.

After opening

Although urine is a liquid, it contains in itself an earthy and adustious part, or like the chemist's tartar and salt, like lime, from which material a calculus arises. This [calculus] is very often found. Sometimes between the tunics so that it can not be removed. The same material or tartar is in the urine of all like lime; [like] *coperas*[499] *or Alome* water, *heat and putt into a vessell will candy;* wherefore in all urines they eject *such a gravelly furr and crust all pissing corners* [which are] *yeallow.*

NB

Wherefore in serpents, birds a double material and the thick surpasses the thin WH

A calculus is generated in the same way as vitriol. *Alumes*, saltpeter &c. and [the way] in which *sugar candy* [is made] and *as will fur the vessel with tartar; att Wether feild:*[500] *stony spring.* Wherefore those of whom the nature is such, by such material in their urine a calculus is created and grows easily and swiftly.

I believe the sediment from the same special material [is important] because where [the calculus is] white, smooth, consistent, with acuminate shape, there [is] no gravel, and where *store of gravell* there it is either lacking or more depressed with many particles and formless slime *sticking to the side.* And WH Δ as the sediment drops down [there is] a larger [amount of] gravel and *falling to the bottom;* on the other hand, the sediment greater &c. as particularized [particles] and suspended. In certain persons not only gravelly material but calculi in the bladder and kidneys and sometimes in the veins. In the liver. Colombo in regard to [St.] Ignatius, General of the Jesuits; in the bladder, ureter, kidneys, colon and in the haemorrhoidal and even umbilical veins and gall bladder.[501]

55r

Substance [of the bladder is] membranous, white and sinewy for strength and retention. Wherefore if wounded

---

[499] Coperas or copperas, meaning ferrous sulphate or green copperas, used for dyeing and tanning.

[500] Wethersfield, a village in Essex about 6 miles northwest of Braintree.

[501] Colombo, p. 491: "In the venerable Ignatius, General of the Society of Jesus, I saw little stones in the ureters, bladder, colon, haemorrhoidal and umbilical veins as well as in the gall bladder."

it does not consolidate[502] except in the neck, as we have seen daily in dissection; especially in children, *torn not cutt*. The internal surface very smooth, *slippery*. WH I saw it ulcerated; and because sinewy, *slimy waters, as vlcers of the Joynts;* wherefore much white, aqueous urine, *as whit of egg*. WH I saw it ulcerated in lues venerea through the whole internal region for years, the kidney intact; the thick, fleshy bladder, as a matrix for it, internally *like vnshorne velvet livid* gangrenous with fetid, disturbed and purulent urines.

[It has] three membranous tunics, one common, two 55v special. The common exterior one from the peritoneum; thin, strong, dense where it is connected to the surrounding parts, viz., rectum intestine, to the ischial and pubic bones; *yett in a woman* I WH saw a prolapsed uterus and cured it. The two special [tunics] for the sake of security and strength, [are] each thick, solid and tough. The interior [tunic is] transparent, very white, thin, sinewy, [with] fibers of all kinds, as nowhere [described by] the famous Vesalius.[503] Transverse [fibers outermost], oblique in the middle [and straight innermost]. When children play they recognize this.[504]

The fundus [is] a little wrinkled, mucous and crusted, [and is] fortified against the sharpness of the urine. The exterior [membrane] thicker in wrinkles, whitish fibers like those of the stomach and intestines;[505] these WH in inflammations [appear] almost as if a fleshy membrane. By these fibers it contracts itself, retains and expels the urine.

The insertion of the ureters [is] into these tunics; where they are thicker near the neck; [they run] obliquely and twistedly between the two tunics, the width of the little finger internally between ingress and egress; therefore

[502] Cf. n. 215.
[503] Harvey may have been thinking of Vesalius's remark (*Fabrica* [1543], p. 518): "for in no tunic of the body [of the bladder] does this triple kind of fibers occur." If so, since the remark was deleted from the revised edition of 1555, this would indicate that on at least one occasion he employed the first edition of the *Fabrica*, since otherwise it appears that he probably used the later, revised edition; cf. n. 517.
[504] *Fabrica*, p. 444.
[505] The muscular coat of the bladder.

when the interior tunic is dilated, it compresses in the form of a little valve[506] lest the urine return; therefore lest distended it be burst by flatus. WH The entrance is so near the neck that the bladder appears like a diverticulum, and it is the reason why flatus does not pass through; for it is shut off.

The neck of the bladder *a little torning* from the interior [runs] from the pubic bone to the root of the penis and thence to the tip of the penis; thus a straight catheter is introduced with very great difficulty. Here occurs section of the perineum for calculus. In women [it is] shorter, wider, extending straight downward, and ends above the opening of the vulva.

56r

Muscles are in the shape of a sigma 𝒮 [and] from their use [are called] sphincters; fleshy fibers, partly transverse partly straight, above the gland called the prostate and at the beginning of the bladder where the sperm issues; lest the urine flow out. WH Indeed, the tensed penis can emit [urine only] with difficulty because the dilated prostate [gland] compresses. Wherefore when this muscle [is] loosened, [there is] a flowing forth of urine; if weakened, incontinence of urine. Therefore after section for calculus, involuntary micturation [frequently results].

To see the transverse muscle lying hidden here, boil the bladder a little; [it lies] between the straight fibers of the exterior tunic.

A muscle X is present for the sphincter;[508] beyond the [prostate] gland [where] you will see fleshy transverse fibers girdling about the canal. But [if it were to relax] the urine would flow forth with the semen. WH Yet I [believe] also to serve the excretion of urine, wherefore *the last of the vrin spirted owte.*

[According to] Riolan the sigmoid valve in the orifice.[507] WH Beware the introduction of a catheter

The bladder boiled

Bauhin

[506] A good description of the course of the ureters through the bladder wall.

[507] Riolan, p. 247: "In the urethrae on the caruncle closing the extremities of the ejaculatory vessels, there is a little membrane at the orifice of the neck of the bladder which prevents reflux of semen into the bladder."

[508] The sphincter urethrae muscle surrounds the membranous urethra just below the prostate as Harvey describes. He is quite right when he says this muscle is "also to serve the excretion of urine, wherefore the last of *the vrin spirted owte.*" for this is precisely its action.

141

The passage in the penis [from] the bladder [is] pro- 56v
longed to the end of the rod, *or rather* the ureters [are] prolonged; diverticulum of the bladder, wherefore OURE-
THRA. Situation under the bodies of the penis, in the
middle; downward from the gland [which is below] the
bladder.

Urine   The pathway [is] common [to both] the urine and sperm
but Vesalius [writes that] a youthful legist from Forlì had
a double canal, one for urine the other for sperm.[509] The
Arabs [describe] two passages, which perhaps [occurred]
commonly at the time because they micturate rarely. WH
I believe one passage to be better so that sperm from coitus
subject to corruption and perhaps corrupted, may be
flushed out; wherefore the Turks often cleanse [them-
selves] like the Venetian prostitutes lest they be befouled.

Sperm   Longer passage for sperm in man than in woman, from
which longer and greater pleasure; like a gullet; wherefore
the principal part, the canal of the penis, and the rest of the
parts [are designed] for the sake of this. A little broader
in the glans where purulent matter is collected in coitus
and the matter of gonorrhea remains. It is turned back here
and here ℔ where the matter remains, and it adheres
through stickiness; wherefore in these three parts, viz.,
in the end of the glans and here and here, ulcers and pain.

Sensitivity   For ulcers or putrefaction [occur] with the greatest pain
because of the keenest sensitivity; wherefore the greatest
pleasure in coitus, and stiffness from micturation. It ulcer-
ates; long *scratches* such as *in a Rat*, from glass; wherefore
when cured the material dries [in] *long thridds*.

The passage of the penis [runs] through the lower part 57r
[of the perineum] to the neck of the bladder. The passage
for the sperm [is] obscure and is found with difficulty,[510]

[509] *Fabrica*, p. 454: "Here at Padua is a young man of the noble family of
Symione of Forlì, a friend of mine and a law student, who has two passages
at the tip of the glans, one for semen and the other for urine." Vesalius errs
when he says one of the openings was for semen, the other for urine. This
would be a case of a double opening in the glans, but there would have been
only a single urethra.

[510] The openings of the ejaculatory ducts are hard to see; they open on the
colliculus seminalis, which is an elevation in the middle of the prostatic portion of the urethra.

142

because it is closed so that it escapes sight except in gonorrhea or at the time of coitus.

WH I saw an ulcerated fistulous *eylet hole* in gonorrhea
with hard lips as the mouth of the fistula; wherefore afterward I cured certain ones of old gonorrhea by the use of
catheters.

NB. Here a caruncle [is present] which is easily in-
flammed; wherefore by unskillful use of a syringe inflammation and pains and especially something of melancholia;
and with poor constitution lethal *gangreynes*. Just as a
reading from [Laelius] a Fonte, p. 588.[511] Sometimes
spongy *prowd flesh swell* so that ischuria may follow.

*It is easyly felt and as easyly Inflamed: be tender and soare
by vsing searing candle and streyning bleed of every little
touch.* Hence *the terrible dissease* caruncle[512] *but many* are
deceived by this which [is] natural, *yett sometime swell as
Sir Thomas Hardes.*[513] WH Pressed through imperceptible
porosities the sperm goes forth, or at least a flavescent
humidity from the glands.

Twisted in the shape of varices located near the deferentia; like first conception of hen's eggs, *ratt, like a little
cocks-combs;* connected to the bladder, wherefore certain
ones with pleasure have ejaculated sperm into the bladder
through diapedesis and have brought forth children; where-

57v

Vesicles

Shape

[According to]
Riolan the
vesicles
compressed in
horseback
riding a cause
of sterility
among the
Scythians[514]

[511] Here Harvey's reference to the page number indicates clearly through
comparison that this is a reference to the *Consultationes Medicinales* (2d ed.;
Frankfurt, 1609), p. 588: "I realize that by a badly administered syringe inflammation is aroused in the neck of the bladder, thence suppression and
fever, which killed a very distinguished man in four days. When the cadaver was dissected there was an ulcer and inflammation in the neck of the
bladder."

[512] *Caruncle* means literally a small fleshy excrescence, but was formerly
applied to urethral stricture.

[513] Sir Thomas Hardes, or Hardres (d. 1628), incorrectly given as Hardy
in the published transcription, was of an old family whose seat was at
Hardres Court, Upper Hardres, near Canterbury in Kent. He married
Eleanor, sole surviving daughter and heiress of Henry Thoresby of Thoresby, a master in chancery. His eldest son Richard, born in 1606, succeeded
his father on March 29, 1628, and was made a baronet on June 3, 1642. His
fourth son, also Sir Thomas, became M.P. for Canterbury and King's
serjeant-at-law. See G. E. Cokayne, *The complete peerage* (London, 1887–
1898), ii, 178.

[514] Riolan, p. 272: "From the site of the seminal vesicles one can understand the sterility and impotence of the Scythians, caused by the aridity of
their seminal vesicles from constant horseback riding."

143

fore in all [there may be] turbid urines after coitus. Use: for collecting semen; like the receptacles for other excrementa so that in the time of coitus it may be ready.

According to Aristotle castrated bulls have procreated;[515] wherefore all copulate so many times according to their number; probably it is 6 or 8. *some lusty Laurenc will crack &c. 12 times but few pas 3 in one night or it is possible can gett above [i] or ij with Child* in one night. *Example* of Turk: perhaps to perfect sperm; for not simply testes make productive but by descent of spirit; and by heat. Wherefore 2nd or 3rd perhaps productive just as of boys and old men; wherefore sperm of women not fecund but unproductive as of boys &c.

Many things are said of number; X lest they putrefy but with great difficulty, perhaps lest they concrete by heat; wherefore in voiding it issues in drops; without venery one coction *great purs: Moul May 3*, or lest all issue in one coitus.

[According to] Varolio if you compress either the smaller testes or the prostate, nothing issues forth, but [if] at the same time [you compress] the vesicles and prostate, the material contained here issues through the canal in the form of milk;[516] I think sperm since the descending spirit causes it to froth; wherefore [sea-borne] Venus, Aphrodite.

This kind of material becomes flavus at the testes of woman &c. [According to] Vesalius that kind a cause of hysteria;[517] WH emission in coitus.

---

[515] *H.A.*, 510a:4: "And I may here add that a bull has been known to serve a cow immediately after castration, and actually to impregnate her." *H.A.*, 632a:20: "If a full-grown bull be mutilated, he can still to all appearances unite sexually with the cow."

[516] Varolio, *Anatomiae* (Frankfurt, 1591), p. 88.

[517] *Fabrica*, pp. 459–460: "In addition to vessels, the ovaries of women have internally certain sinuses filled with a thin and watery humor. At first in dissection, an uninjured ovary when compressed, like an inflated bladder, usually with a crackling noise squirts forth this humor to an astonishing height, not unlike a fountain. As that humor, like very thick whey is white in healthy women, so also I have found it yellowish like egg-yolk and a little thicker in two girls of high nobility, who before death suffered from strangulation of the uterus. In them merely one of the sinuses of one ovary, stuffed with that yellowish humor, protruded like a somewhat large pea, and

Nevertheless Bauhin and others believe on the contrary that [there is] a humor here [and] that it coats the passage lest [there be] sharpness.[518] WH On the contrary, for [there are] two humidities.

1st, because *Ratts and monkeys* are divided; *Moule* inward, these outward; prostate gland more externally. Internally *like a cockcomb* full of sperm; prostate gland in female *Ratt* so large that I believed it for testicles and to be a hermaphrodite. Therefore because very similar to other glands, I believe the use of the other glands to be to receive humidity *to supple slipper and defend* from sharpness.

2nd, the material itself in the vesicles; the same condition as that in which sperm dropped cold on earth; here, before the arrival of spirit; there, the same [spirit] wholly vanishes which true gonorrhea.

In the glands frequently a very clear water [that may] counterfeit gonorrhea, especially in the lustful; wherefore [according to] Galen by tickling to arouse sexual desire;[519] wherefore its humors arouse the lustful to lechery; very often they are seen to touch and to tickle the genitalia. I believe this humor the cause of satyriasis, and with Rondelet *pisse* chaude.[520] *Burning* either having been consumed or corrupted.

Question whether in women [there is] another sperm; that [according to] Aristotle a special argument against those who [declare] that sperm affects women with lecher-

---

tinting the adjoining parts as we observe the colon in men, where it is carried under the liver, to be rendered yellowish from the gall bladder. As the color of this humor or liquor occurs rarely, so also it had a very pronounced odor and revealed itself as offensive, the poison and undoubted author of those symptoms of the brain which daily arise in strangulation of the uterus." This material is not be be found in the 1543 edition of the *Fabrica*, thus indicating Harvey's usage of the second edition of 1555 or one of the reprints of it.

[518] Bauhin, p. 191: "In the past we have observed that these vesicles are full of an oily, yellowish humor, . . . by which the common passage for semen and urine is coated lest it be harmed by the sharpness of the semen or urine, and lest dried out it settle and prevent the urine and semen from being carried easily through the lubricated channel."

[519] *De Usu Partium*, XIV, ix; Kühn, IV, 180; Daremberg, II, 113.

[520] *Gulielmi Rondeletii Methodus curandorum omnium morborum corporis humani*, (1609), pp. 536, 548.

ousness.[521] Indeed, in furor uteri in women it becomes putrefied and smells bad; and from motion and agitation there is caused a very white froth and [produces] sordes as in men; frequently also it falsely appears to be gonorrhea and tumors of a white flux.

That odor an indication of lustfulness in animals and of healthiness in women.

It causes sordes in all, wherefore perhaps psylotra[522] and 58v circumcision [are required]; in use and odor it is similar to the mucous of the intestines.

Site of the prostate [is] at the neck of the bladder; two glands beyond the muscles where the deferentia are joined as in the testes. WH Perhaps the testes of women [are] analogous to these to which [they are] more similar than to the testicles. Without these Rufus says that eunuchs emit a certain semen from glandular fistulas,[523] but infecund, as I believe in women WH.

The membrane by which the deferentia are covered [is] dense and very thin. Many passages and blind little spirals from which humidity like quicksilver is expressed from the corium in drops, viz., compressed like blisters. All the passages of the genitalia [become] lax in gonorrhea; see Vesalius.[524] Here WH in certain ones I have cured many calloused, fistulous ulcers, from which pus [goes] into the bladder and into the anus if the ulcers [are] corrosive, creeping ones. Here a large ulcer in *a friar* so that dead from a wasting marasmus, WH Δ. Very old, little ulcers with discharge of pus through the penis. White, smooth and even. [According to] Nicolò Massa[525] these having

    [521] *G.A.*, 727b:34: "Some think that the female contributes semen in coitus because the pleasure she experiences is sometimes similar to that of the male."

    [522] The meaning of this is not quite clear. Psilothra are preparations that remove hair. Psilothrum is another name for *Bryonia alba*, the white jalap, the dried root of which is both a vesicant and a purgative.

    [523] Rufus of Ephesus, *De Appellationibus Partium Corporis Humani*, II, xiv, in *Medicae Artis Principes post Hippocratem et Galenum* (Geneva, 1567), coll. 119.

    [524] *Fabrica*, p. 451: "In those suffering from an involuntary flow of putrid semen, in the first days, along with a loose and flaccid penis, they suffer also from an itchiness."

    [525] Massa, f. 34r.

been removed, the burning heat of urine [has been cured]. WH The worst burning without discharge of pus; by dryness; just as the tongue is inflamed without humor, *chapped: sore.*

[There are] various names of the mentula; penis from 59r pendant; a rod. 20 Greek names, 16 Latin, according to [Du] Laurens;[526] indication of fecundity by giving birth to so many names for itself.

Site convenient for introduction of sperm into vulva. In the middle of the body because without a mate; in forward part of pubic bone; outward when swollen as appears in horses, but retracted inward [when] without sexual desire. WH Retracted strongly in suppression of urine, dropsy and hernia. Retracted, colder so that certain ones believe man can decline into hermaphrodite or woman.

WH in the case of a boy at Padua by the bite of dogs; afterward retracted, as they say is usual in eunuchs; *like a monkey ij stones and noe yearde.* [Not always] distended; only during and before coitus; for otherwise an impediment, like a hand always stretched out tensed. For by tension itself the spirits are destroyed; very weak body so that bodies become weakened easily by priapism; they are exhausted and gradually consumed; for turgid by spirit.

Shape noted most in all $\mathcal{U}$, differs in length and thickness in men and animals; in length is proportionate to the vagina of the uterus, so also [proportionate in] thickness and straightness.

WH I believe a thick, rounded shape is preferable to a long one; hot [sperm] because [according to] Aristotle in those in whom long too much infecundity, [even if] *robust,* because the sperm is cooled [IV], p. 613.[527] On the other hand, physicians [consider] shortness of the penis as a disease.

[According to] Galen greater coitus [after] abstinence, 59v De locis [affectis], VI, 6.[528] On the other hand, WH sculp-

[526] Du Laurens, p. 262.
[527] *G.A.,* 718a:23: "This happens also if the penis is large; such men are less fertile than when it is smaller because the semen, if cold, is not generative, and that which is carried too far is cooled."
[528] Galen, *De Locis Affectis,* VI, 6; Kühn, III, 450; Daremberg, II, 704.

tors present it as retracted in statues of athletes, and in the aged it is cold and infirm. NB **WH** an equilibrium: as much as the testes hang down, the virile member is retracted. It is augmented in boys caressing girls; it would be longer if the midwives [did not] cut the umbilicus [close].[529] Whether **WH** because it is drawn inward with the bladder, urachos. Erect it is moved and strikes the belly as by micturation of a horse; spirits having been driven into it; *blow a glove.*

Action: to prevail strongly and fiercely over women through the sensation of tickling; to entice to procreation and to ejaculate into the uterus; wherefore in those animals which introduce female spermata, *Snayles Cockles*, place one exactly against the other, and they introduce a very viscous sperm; wherefore in some animals it is not because they introduce a particle *in mas* of the uterus. Silkworms, butterflies; and they constitute the heat of the body.

It is divided primarily into the superior part SĒMA, which is erected; the inferior UPOSĒMA *yeard prick pintle and rote of the yard.*

Division: parts are skin, epidermis and fleshy or sinewy membrane. The flesh within [is] blackish and spongy; glans and its membrane; veins, arteries, nerves, muscles. The canal of which [mention has been made] before **WH**.

Division into right and left, for [it is] double and in two bodies.

The skin [is] thin, lax, flaccid without fat; *swelled like* 6or *pudding* in dropsy; in all regions some are circumcised in accordance with their religion;[530] **WH** I believe because of leprosy, because in hot regions much filth, wherefore

---

[529] Crooke, p. 210: "It is thought also it wil be longer if the Nauell-strings bee not close knit by the Midwife when the childe is new borne; and that because of a Ligament which commeth to the Nauell from the bottome of the bladder which they call the *Vrachos*, for the straighter that is tyed vnto the Nauell, the more the bladder and the partes adioyning are drawne vpward." This theory persists in popular belief even today. In Harvey's time it was obviously firmly believed.

[530] The published transcription is incomplete. Harvey's complete statement is *praeputium aliquibus regionibus omnibus religione circumcisum* and has so been translated.

cleansing by the Turks and by the Venetian prostitutes.

Some circumcised by nature; the manner of circumcision; on the contrary, in others the glans is never uncovered; phimosis, a little cohering to the glans [is] sometimes *vncuttable*.

In some a frenum; sign of virginity; paraphimosis, comparable to the hymen of virgins. The Romans [practised pinning] lest the slaves constrain the women to venery. Method of pinning, see Celsus, p. 667.[531]

WH The circumcised have less pleasure in coitus because the membrane is thickened; sensitivity blunted.

The circumference of the glans, exceeding the trunk, [is] called the corona; very keen sensitivity, where sperm delights, and in children and the delicate [it is] very soft and rare; wherefore it produces greater pleasure, and is more swiftly injured by impure venery.

WH I know a Paduan capable of coitus with the glans removed.

WH Compressed [in erection] it becomes white, then red, inflated; a very thin membrane continuous in the internal canal.

WH in the seal very long and rounded *like a* nasal polypus; leather pipe.

Membrane fleshy, sinewy, and very strong lest the spirit be swiftly dispersed.

In dogs, wolf, weasel, *giney cuneys*, osseous or the whole of their mentula osseous. **WH** *a pretty bable lord Cary*[532] *a*

[531] Celsus, *De Re Medicina*, VII, 25; W. G. Spencer trans. (London, 1953), III, 423–425: "Some have been accustomed to pin up the prepuce in adolescence either for the sake of the voice, or for health's sake. This is the method: the foreskin covering the glans is stretched forwards and the point for perforation marked on each side with ink. Then the foreskin is let go. If the marks are drawn back over the glans too much has been included, and the marks should be placed further forward. If the glans is clear of them, their position is suitable for the pinning. Then the foreskin is transfixed at the marks by a threaded needle, and the ends of this thread are knotted together. Each day the thread is moved until the edges of the perforations have cicatrized. When this is assured the thread is withdrawn and a fibula inserted, and the lighter this is the better. But this operation is more often superfluous than necessary." Harvey's reference to page 667 permits us to determine that he employed the edition published at Leiden, *ex officina Plantiniana*, 1592.

[532] I.e., Robert Carey, first Earl of Monmouth, 1560?–1639.

whale as big as his middle. In man sinewy, hollow [like] a leather pipe.

[The body of the penis is] bipartite right and left; they 6ov arise from the lower part of the pubic bone and the upper [part] of the coxendix[533] for stability. Below at the origin they are divided like two fingers, thus   so that space may be given, room for a canal; the bladder.

Inwardly [filled] with spongy flesh for inflating with spirit, for receiving as water by a sponge, similar to a blackish spleen. The substance of the canal [is] also of this sort and also the vagina of the female uterus. NB WH here spirit distending into priapism. WH Perhaps in the spongy, blackish flesh the genitalia of men as well as of women apt to cancer; wherefore Du Val,[534] barber-surgeon, deceived by blackness of this flesh removed it as a cancer with great pain. Regarding the muscles [see under] the anatomy of muscles WH.

The thorax [is altered] by constant motion or spring. 6ir Use: to serve these contained parts; and [act as] a support for the muscles and scapulae. [According to] Galen four uses:[535] for the sake of the heart, lungs, respiration, voice.

Its shape in man [is] *more flatter* than in other animals. In animals a keeled chest [is common] as in the ape, dog, and in certain men *who* [are] *out chested.*[536] Nature supplies it with depth when it cannot have width.

Cause. Why rounded (by avoiding angles [is] less liable to external injuries). Rather it ought to be asked why it is not rounded in other animals, since according to its nature more like trees, with arms and legs to the sky. Therefore in some [is] long in which the bodies [are] long such as

[533] This term was originally applied to the whole of the hipbone. Later it was used to mean only the ischium.
[534] Possibly Jacques Du Val of Evreux who practiced with some distinction in Rouen and was the author of *Methode nouvelle de guérir les catarrhes* (Rouen, 1611) and *Des hermaphrodies* (Rouen, 1612).
[535] *De Usu Partium*, VI, ii; and more exhaustively in *De Usu Respirationis*.
[536] This refers to the deformity produced by rickets, common in Harvey's time, resulting in "pigeon chest." Rickets was first described as a specific disease in 1645 by Daniel Whistler in his graduation thesis at Leyden, *De morbo puerili Anglorum.*

serpents, worm; in others projecting so that the feet are very close together as the *greyhownd*. In birds keeled *as the shipp keele*[537] like a foot *to sitt on*, to rest and so that there be stability for the muscles of the wing which are very large and very useful. In man stability for the muscles of the scapulae and of the arms [and] abdomen.

The size [is] proportionate to the heat; therefore in all animals an indication of the supply of heat, animation, boldness. Wherefore the chest [and] nostrils [are] *ample;* in horses *Brode Bresten;* in falcons, *hawkes;* the emperors [according to] Suetonius [were] of broad shoulders;[538] *greyhound deep chest.*

[According to] physicians [those with] a broad chest 61v easily vomit; [those] with a narrow chest rather than broad [are] pusillanimous, timid. Wherefore [according to] physicians [those with] winglike scapulae prone to tabes.[539] In a tabetic Δ dissected, **WH** the scapulae having been elevated as the flanks retracted, from constricted chest *ther armes hang of and appere longer the chest shrunke from them.*

Where all things coalescing the cavity of the thorax of the capacity of a fist; *sett ther breath one* ⅓ *of the lenght.* Length in man one face, just as from one nipple to the other, from chin to hair.

Question: that a very hot animal is determined by the size of the thorax; sanguineous hotter than nonsanguineous, for heat [comes from] spirit and blood; and sanies [and] ichor in the nonsanguineous, for the blood [is] crude, diluted, ichorous.

Wherefore in the crude blood of woman [are] white fluxes in place of menses; wherefore butterflies in the sum-

[537] Aristotle, *De incessu animalium*, 710a:30: " . . . the strong and acute breast-bone (acute like the prow of a clipper-built vessel, so as to be well-girt, and by strong dint of its mass of flesh) in order to be able to push away the air that beats against it, and that easily and without exhaustion."

[538] Harvey's remark appears not to be based on any such general statement by Suetonius, but rather upon the latter's description of Tiberius as "broad of shoulders and chest."

[539] Tabes is used here to mean any wasting disease. The reference to winglike scapulae is interesting, for this condition follows paralysis of the serratus anterior muscle. Harvey's statement could well be modified to read: those with tabes prone to winglike scapulae.

mer flourishing on a drop of blood Δ WH; those with wider chest [have more blood], wherefore hotter.

Especially in summer the newt is hotter than fish; nor ought it be said that fish rejoice in the torpidity [produced by the summer heat], viz., in the water.

Animals having lungs are hotter than others; for [they have] more blood.

In fish, as in newts, lizards and frogs, the chest is pro- 61r portionately smaller. Among hotter men those who have a greater quantity of blood or more spirit, therefore [have] a more ample chest. Wider nostrils so that they provide both greater space and opportunity for ventilation. In those in whom if the chest is deficient, the heat of the heart causes this part [to be] more ample; in the pubescent Δ as the voice is altered so the chest; and also in one deficient in height, as hunchbacks, depth is supplied. Wherefore many at the time of puberty become hunchbacks before sufficiently erect. Wherefore hunchbacks [are] animated and bold, especially in speech because *ther hart neter ye long*; but WH from a greater supply of heat *more bowld and prompt*.

Motion or action of the thorax in respiration.

Respiration [has] two parts: inspiration and expiration; two ways; three degrees [of motion]: free, violent and very violent. X A[cqua]p[endente]: in three ways. Free [motion] from the diaphragm and underlying parts; violent from the intercostal muscles; very violent from the muscles of the scapulae and abdomen.[540] WH in each way it is free, violent and very violent.

Respiration occurs in those two ways; Δ in the two sides of bellows, now in the one way and now in the other; the upper part in the bellows is where *ye clack*[541]

But two ways in man; I do not see [a third] in others. 62v

[540] Hieronymus Fabricius, *De Respiratione et ejus Instrumenta,* in *Opera Omnia Anatomica & Physiologica* (Leipzig, 1687), pp. 174–183; chs. 9, 10, 11; respectively "De expiratione libera," "De violenta respiratione & ejus instrumentis videlicet musculis intercostalis," and "De violentissima respiratione & ejus musculis."

[541] A "clack" was a kind of valve used on pumps which was opened by the upward movement of the water produced by suction and closed by the backward pressure of the weight of water. According to John Bate, *Mysteries and Arts in foure seuerall Parts* (1635), "a clacke is a piece of Leather nayled ouer any hole hauing a peece of Lead to make it lie close, so that ayre or water in any vessell may thereby bee kept from going out." Cf. Charles Singer, *The Discovery of the Circulation of the Blood* (London, 1922), p. 46.

152

WH In one [way] by inspiration the venter is expanded, by expiration it is contracted. On the other hand, by a second inspiration the venter is contracted, the chest is raised, the scapulae are elevated and the [hypo]chondria dilated. If the hand is applied, the skin and cartilage on each side [are seen] to be raised. [The thorax] is opened almost to the lowest [part].

In the first, inspiration is seen to occur in the diaphragm, expiration by the muscles of the abdomen. [This can be determined] here and there with the two hands; wherefore cold water for hysterical symptoms; extenuation of the venter, by retraction of the hypochondria and flank, difficulty of respiration. Wherefore it appears that respiration [occurs] as if *in the flanke*; in birds in the fundament.

WH How in serpents and birds in which a diaphragm is lacking

In the second way, respiration is seen when the first [venter] is impeded; for the first venter having been distended the second may inspire further; the *belly* having been distended with the *lung*. In hysterics *cutt laces*; likewise in pregnant women; *they all vse this panting sighing fashion of breathing*. Sign of the greatest weakness if the chest is raised by the scapular and pectoral muscles. *Breake buttons*.

Δ Inspiration for each occurs in one way: expansion; as far as possible and then the second is permitted

The thorax [is] divided into parts: *conteyning, conteyned,* attached, partly contained, partly containing. [1] Containing: common, special. Common are: skin, epidermis, membrana carnosa,[542] fat. Special [are]: sternum, cartilages, ribs, clavicles, vertebrae, soft intercostal muscles, veins, arteries, nerves. [2] Attached: muscles of the scap-

63r

---

[542] Riolan, *A Sure Guide*, trans. N. Culpeper (London, 1671), p. 37: "The fleshy membrane lies under the Fat, and sticks to it, and is conspicuous in young Children newly born, where it is not hid with Fat. It is more obscure in such as are grown up, and yet it retains its fleshy substance, as is evident about the Loynes, Cods, Forehead and Neck. . . . In bruits it is next to the Skin, which often moves by the intervening of this Membrane." Gibson, *Anatomy of Human Bodies* (London, 1688), p. 16: "The Carnous Membrane is only properly so called in Brutes, in whom it is truely fleshy and muscular; but in Man it may more properly be called membrana adiposa, or the fatty membrane, according to Dr. Glisson, seeing it has no carnous fibres, or parenchyma, but is a membranous substance stuff'd with some fat." Nicolò Massa, in his *Anathomia* (Venice, 1536), was the first to use the term *panniculus carnosus*, although he greatly exaggerated its extent and importance.

Three conditions of respiration: [1] cavity as it is in the chest for containing air; [2] opening as it is through the nostrils and mouth; [3] dilatation and constriction as of the bones, muscles

ulae, breasts, nipples, emunctory glands,[543] neck, base of neck and its parts. [3] Contained: passage ways and special [parts]. Passage ways: gullet, great artery, ascending vein in the thorax and neck,[544] nerves of the 6th pair, recurrent and of the diaphragm.[545]

Special [parts are those] according to nature, [those] contrary to nature. According to nature are: heart, lungs, thymus, azygous [vein]. Contrary to nature: I have seen water like sanies in the pericardium. Likewise *pus in these two powches* [and] blood from penetrating wounds. Abscess from aposteme in the pleura and tumors. [One] opened at Padua, from which a great quantity of sanies and pus from empyema, by Gulio [Casserio] of Piacenza; *foetid.*[546] Giulio Jasolino each day for 30 days [collected] 3, 4, and 5 measures.[547]

WH In dropsy of the lung [one can get] *above a galon* of ichor; doubt how it may be evacuated through the urine; doubt of paracentesis, some higher some lower because of descent of material; because of diaphragm.

WH Wheresoever matter has fallen into [the lung], evacuation [is] by cough and by reason of the site of the body; a natural demonstration that vital [spirit] for the veins, arteries [and voice is abundant in the region of] the highest rib.[548]

[Structures are] partly contained, partly containing: 63v diaphragm, disseptum,[549] pleura, membrane, arteries, veins, intercostal nerves.

Of the common parts: skin, epidermis, membrana carnosa [there has been] sufficient [discussion] before. Of the muscles of the scapulae, serratus major, [consideration]

[543] The term is here applied to the axillary lymphatic glands.
[544] This refers to the superior vena cava, the innominate veins, and the jugular veins.
[545] The vagus, recurrent laryngeal, and phrenic nerves.
[546] The published transcription provides the meaningless *fistid* for foetid.
[547] Jasolino, *De Aqua in Pericardio Quaestio Tertia* (Naples, 1576), no pagination [B⁴ᵛ]: "for about thirty days I collected each day three, four and sometimes five measures of fluid."
[548] Cf. Crooke, p. 347, and Galen, *De Usu Partium*, VI, xiii; Daremberg, I, 426.
[549] A term formerly used for the diaphragm

among the muscles. The breasts and nipples [will be dis-
cussed] when the female genital parts [are considered];
similarly of the generation of milk; whether it is attracted,
propelled or [moves] of its own accord; whether the same
as menstrua. Likewise whether and how milk is concocted.
The nipples in men [serve] as a distinguishing mark; indication, if it be necessary, where two nerves are present
at the nipples; four others; wherefore sensitivity to both
pain and tickling. The breasts [are] for the sake of milk
wherefore [exist] in those [animals] in which necessary;
regarding the milk of the oviparous.

Breasts [develop] with the arrival of sexual desire, at
which time [girls] are carnally sought by men inflamed
with the onset of boldness; as among the rest of the animals. Glands with all fat vanish in the aged. In [men with]
effeminate constitution the breasts [may be enlarged]; and
in some milk, *Sir Robert Shirley*.[550]

As Aristotle,[551] Albert, [552] Avicenna,[553] Cardan[554] testify
[to this] wherefore [it is] not astonishing if milk in virgins
and the nonpregnant; but WH not milk but that kind [ex-
isting] in false conceptions; something similar to milk and as
much like milk as man's sperm is like the sperm of woman;
wherefore *she hath water in her brest;* but [according to]
Aristotle milk [in] the goat of Lemnos;[555] and in the new

[550] Sir Robert Shirley, 1581?–1628, the famous envoy to the Shah of
Persia. Cases of male lactation are to be found in the medical literature available to Harvey; e.g., Aristotle, Avicenna, Albert, and Cardanus. There are
well-attested modern cases.
[551] *H.A.*, 522a:19: "With some men, after puberty milk can be produced
by squeezing the breasts; cases have been known where on their being sub-
jected to a prolonged milking process a considerable quantity of milk had
been educed."
[552] *De Animalibus*, ed. Stadler (Münster i W., 1916), III, 170: "Sometimes non-pregnant women of superfluous humidity will have some milk in
their breasts, and especially virgins and those who have abstained from intercourse for a long time."
[553] *Canon*, Lib. III, fen. xii, cap. 3: "And sometimes milk collects in the
bodies of men, and especially the pubescent, so that their breasts are
rounded."
[554] *De Subtilitate*, bk. 12, in *Opera Omnia* (Lyons, 1663), III, 557: "I saw
. . . Antonio Benci . . . thirty years old, from whose breasts there flowed
so much milk that he might have suckled a child."
[555] *H.A.*, 522a:13: "once in Lemnos a he-goat was milked by its dugs
(for it has, by the way, two dugs close to the penis), and was milked to such

world.[556] *Lady Hervey*[557] milk was harmfully withdrawn from excoriations &c.; elsewhere.

The neck for adornment; for all are adorned. Reason for 64r the pharynx [according to] Galen, [De] U[su] P[artium], VIII, [ch.] 5; and by pharynx the voice wherefore [according to] Galen nothing lacking a neck produces a voice.[558] WH But X the rucoli[559] of frogs. WH *make frogg squeake. A porpos* WH so that not produced in those in which neither neck nor lungs; rather [according to] Aristotle by reason of lungs which are lacking without a neck;[560] or by reason of the trachea so that the air be somewhat cooled, wherefore by uvula to temper the air; it draws away ptisis[561] from the lung.

Neck of birds used for hand, like proboscis of elephant; wherefore [there is appropriate] length *to reach every part and the communicable* ground; thus those which [have] long legs [have] a long neck; and in those which [have] short legs, if [they obtain] aliment from the ground level, a tongue.

Long neck in men a sign of timidity; if *subtile*, of weakness, as some stags [have] a long [neck], viz., 7 vertebrae, others 5, others 8. Short and thick [neck signifies] steadfastness *hedstrong stubborn;* stiff necked.

Divided: anterior part, root of neck, posterior [part] of

---

effect that cheese was made of the produce, and the same phenomenon was repeated in a male of its own begetting. Such occurrences, however, are regarded as supernatural and fraught with omen as to futurity, and in point of fact when the Lemnian owner of the animal inquired of the oracle, the god informed him that the portent foreshadowed the acquisition of a fortune."

[556] Crooke, p. 194: "They that haue trauailed into the new world do report that almost all the men haue great quantity of Milke in their breasts."

[557] Born Cordell, youngest daughter of Brian Ansley of Lee, Kent, on February 5, 1607. She married William, Baron Hervey of Kidbrooke, Kent, to whom she presented three sons and three daughters. Lady Hervey died May 5, 1636.

[558] *De Usu Partium,* VIII, i; Kühn, III, 609–610; Daremberg, I, 524–525.

[559] A word of Scandinavian origin preserved in later dialect forms as *ruckle,* to make a rattling or gurgling sound in the throat. Lyndesay in 1530 wrote of "a ruclande raven."

[560] *P.A.,* 664a:20: " . . . the larynx, exists for the sake of respiration, being the instrument by which such animals as breathe inhale and discharge the air. Therefore it is, when there is no lung, there is also no neck."

[561] Ptisis was a term formerly used for phthisis.

neck. Common containing parts: skin, epidermis, membrane, fat. Partly containing, partly contained: various muscles, vertebrae.

Special contained: external, internal jugular veins, carotid arteries; nerves of 6th pair; recurrent nerves; trachea; esophagus; spinal medulla.

Furthermore, mouth and its parts: teeth, jaw, tongue, gums, palate, muscles, vessels.

Fauces: uvula, those tonsils, *Almondes*, epiglottis, glottis, larynx and its parts, hyoid bone.

The jugulars[562] [arise] from the ascending vena cava 64v which [extends] through the mediastinum upward to the clavicles; it is divided at the thymus,[563] sometimes at the pericardium itself. [The vena cava opens] into the heart just as into a cistern;[564] by losing continuity it becomes lax and declines into the ventricle. Internally here [there are] four branches: phrenic, coronary, azygous, intercostal.

The intercostals [join] the azygous [vein] wherefore violent phlebotomy in all pleurisies at the right [elbow] by KAT'IXIN.[565] At the thymus under the pectoral bone there are two branches called subclavian; but axillaries when they go forth from the cavity. From the subclavian: mammary, mediastinal, small cervical, muscular. From the axillary, the scapular, internal and external, viz., in the inferior part, in the superior part; muscular superior in the neck, external jugular, internal jugular and apoplectica.[566]

Artery [1] descending branches: single, cuticular to the back; phrenic, to the muscles of the thorax; [2] ascending:

[562] Harvey applies this term to the innominate as well as to the jugular veins.

[563] The thymus is, more correctly, in front of the two innominate veins. Because of the more nearly vertical direction of the right innominate vein, the thymus overlies the left vein over a considerable portion of its length. Most of the thymus is placed inferior to the union of the innominate veins and thus in front of the pericardium.

[564] The superior vena cava opens directly by a wide opening into the right atrium of the heart.

[565] The use of venesection to withdraw a redundant humor from an unaffected region of the body but on the same side as the ailment produced by the humor.

[566] The medieval anatomists called the jugular veins the venae apoplecticae or venae guidem. It is not clear what Harvey means here, for he refers to the jugular veins by their usual name.

in all the upper parts; subclavian which after its egress [from the thorax is called] the axillary, from which into 3 [branches] in the superior ribs; mammary from the superior part; cervical; muscular.

From the axillary: superior thoracic to the pectoral muscles; inferior thoracic to the whole side; scapular; humeraria to the muscles of the shoulder. Near the higher part of the sternum supported by the thymus; as yet in the cavity of the thorax into two unequal branches called carotids.[567]

A nerve of the 6th pair[568] descends between the jugular 65r vein and the carotid artery; all of which are covered by one membrane at the side of the windpipe under the muscles common to the larynx and sternohyoid.

At the clavicle [it is divided] into two branches, exterior and interior. Exterior to the muscles extended to the sternum and clavicle; the recurrent into the exterior covering of the windpipe. The nerve is turned back [1] at the right above the axillary artery into three little branches, more rarely as one; [2] at the left above the aorta, rearward to the spine between the aorta and the pulmonary artery. Before the said branches, little branches [go] to the heart, pericardium, and covering of the lungs, and [are] attached to the esophagus at the mouth of the stomach.

Interior [branch], the costalis, under the pleura to the vertebrae through the roots of the ribs. From this, intercostal branches [arise]; from these, connections with the nerves from the vertebrae of the spine; it seeks the lower venter with the great artery as far as the sacral bone, bladder, and fundament under the PSOAI muscles.[569]

[567] Here Harvey is confused, for on the right side the common carotid and subclavian arteries arise from the innominate artery. On the left side they arise from the arch of the aorta. It is true that the left common carotid has a somewhat longer intrathoracic course than has the right.

[568] This refers to the vagus or tenth cranial nerve.

[569] This paragraph refers to the sympathetic nerves of the thorax and abdomen. In Harvey's time these were not regarded as being part of a separate system, but were thought to be branches of the "6th pair." The thoracic and abdominal sympathetic chain had been illustrated most accurately by Bartolomeus Eustachius, but as his plates were not published until 1714 they were not available to Harvey.

Seven pairs of cervical nerves,[570] viz., as many as the vertebrae.

Twelve pairs of thoracic nerves called intercostal because in the spinal medulla they are joined, with the internal veins, with the nerve of the 6th pair for strength.

The nerves of the 6th pair with the great artery perforate the diaphragm.[571] It is divided; a right branch with 3 [further branches]: 1st, to the lower omentum in three branches; to the colon, little intestinal branch; 2nd, to the right kidney whence come the gastric and nephritic; 3rd, mesenteric, then straight to the bladder. The left [branch also divided into] 3: 1st, to the omentum and colon from the stomach; 2nd, to the left mesentery and intestine; 3rd, to the left kidney.

The rest into the left [side] of the neck of the bladder under the PSOAI.

In antiquity the trachea [was] called simply the pipe[572] 65v of the lungs; *wind pipe*. It originates at the larynx from which [it proceeds] downward; [also] from the region of the nasal foramina, wherefore whenever liquid is drunk, [air] is inspired through the nostrils, [and one can] breathe forth [through the nostrils] as *tabacco*. It ends at the lungs where it is divided into cartilaginous bronchi [which go] to the extremity of the lung. It divides between the pulmonary vein and the pulmonary artery with which [there is] anastomosis[573] by which sooty vapors from the heart and pituitous material [are expelled from the body].

To the rear of it [lies] the esophagus, wherefore use of

[570] There are in fact eight pairs of cervical nerves. The first emerges above the first cervical vertebra; the eighth below the seventh cervical vertebra.

[571] The vagus nerves pass through the diaphragm with the oesophagus. The sympathetic nerves pierce the lumbar origin of the diaphragm. No nerves go through with the aorta (the great artery).

[572] The published transcription incorrectly gives *camina* instead of *canna*.

[573] At the time he wrote this, Harvey apparently believed some sort of anastomosis existed between the trachea and the pulmonary artery. Later, he would have refuted vigorously this curious survival of ancient teaching. He would probably have used such terms as those he expressed in *De Motu Cordis* in disposing of the fallacy of the pores in the interventricular septum of the heart—"But, damme, there are no pores and it is not possible to show such."

159

the epiglottis, viz., in viviparous [animals] of which the skin is covered by hair; in others the pharynx is sufficiently closed per se.

The esophagus [goes] from the mouth into the stomach [and runs] straight downward to the fifth thoracic vertebra, then yielding place to the great artery. Inclination hence to the right; by the strong membranes of the left [it is] somewhat elevated (lest by the passage of food it become⁀an obstacle to the artery); descending with the aorta into the upper part of the stomach, viz., the left orifice, it ends with two nerves from the 6th pair.

The mouth with its parts as a site [is] well enough known to all.

Fauces, posterior and inferior, part of the mouth; those parts common to the throat and esophagus; where the nasal foramina [are] for things descending into the mouth, by which *Tabacco* ascends.

Here [are] the uvula, tonsils, epiglottis and root of 66r tongue.

Syphon from the nares⁵⁷⁴ Uvula [arises] from the deeper part of the palate; near the nasal passages it hangs down into the mouth; its substance [is] fleshy, spongy, roundish; when swollen by catarrh it is relaxed; [it is] the plectrum of the voice; and tempers the air.

The tonsils, almonds, antiades [are] glandular, spongy [and lie] near the root of the tongue before the beginning of the esophagus. It serves the saliva of the mouth.

Here many ulcers [occur] in lues venerea. In [mercurial] inunction all the material issues here and sometimes corrodes, wherefore the masseter muscles are contracted by scarring.

The epiglottis, compressed by the weight of food, covers the fissure; it is situated at the root of the tongue.

The 3rd pair of nerves [run] to the tongue and lower jaw.⁵⁷⁵

---

⁵⁷⁴ This has not been included in the published transcription.

⁵⁷⁵ The third and fourth pairs of nerves in the medieval terminology, still used in the early part of the seventeenth century, correspond to the modern trigeminal or fifth cranial nerve. The lingual and inferior dental branches of

[The epiglottis is] made from cartilages so that [it is] 66v
hard and dense; not frangible; so that it is free from external or internal injuries through the esophagus; according to
A[cqua]p[endente] so that lying open it may not fall shut
[and block] the air which always and continuously must
enter and go forth; if of heavy bones [it would require
large muscles according to] A[cqua]p[endente], and [if of
light bones] would be easily broken;[576] therefore not osseous; it would impede the transit of food through the
esophagus.[577]

However, [according to] Colombo in a number of dissections of the aged it was osseous;[578] but [according to]
Fallopius [so are] the first and second [laryngeal carti-
lages] never the third and fourth;[579] WH viz., arytenoid.
WH Annular [made] firm by the remaining three; for opening and closing; Δ in the elbow.

By the common muscles WH *lift vp the sack.*

By special muscles because with [special] use: to inspire, to expire, and in each case to inhibit motion; to
breathe out, to cough, to cry out; it is advantageous for
this to be voluntary, wherefore muscles, organs [have] voluntary motion.

---

the mandibular division of the trigeminal nerve subserve sensation in the
tongue and teeth of the lower jaw respectively.

[576] Hieronymus Fabricius, *De Gula, Ventriculo, Intestinis,* in *Opera Omnia
Anatomica & Physiologica* (Leipzig, 1687), p. 102.

[577] Here Harvey is referring to the epiglottis. In the next paragraph he
refers to the laryngeal cartilages as first and second (presumably the thyroid
and cricoid cartilages), and third and fourth (presumably the two arytenoid cartilages).

[578] Colombo, p. 184: "But if you inspect the human [cadaver], especially
in one of more advanced years, undoubtedly you will recognize all the os-
sicles from which it is formed." This appears to be the only remark by
Colombo that bears any relationship to Harvey's statement.

[579] Fallopius, f. 43ʳˑᵛ: "The larynx is placed below the above-mentioned
bone and consists of four cartilages. The first is shield-like [thyroid], being
so-called by anatomists, the second, which has resemblance to a ring, can be
called the KRIKOEIDĒS although it has been dismissed by anatomists
without a name; the third consists of two parts which anatomists count as
only one and call the arytenoid. Consequently there are four. Of the aforementioned I have sometimes found the first and second to have become osseous in the very old, and not only in the decrepit but also in those at the
beginning of old age. However, I have never seen the third, or to speak more
correctly, the third and fourth, osseous."

161

And so nerves for the sake of motion, veins for nutrition, arteries for life, membrane for protection and connection, glands for moisture.

See more in the book of the head **WH**.

The larynx [is] composed of cartilage and muscles [and 67r is] placed in the middle in line with the nostrils. Especially the so-called rima glottidis which is a fistula of the tongue connected at the rear to the esophagus. Prominently forward [at] the Adam's apple.[580] The sign of the *Couckould.* **WH** Perhaps because in a very long and rounded neck, the sign of a cold brain and of a temperament which is impotent for venery and timid; *which two conditions the one makes them wonder the other dare to committ.*

At the fourth thoracic vertebra [the trachea] is then divided into two bronchi; more prominent in man than in woman; **WH** because of the bronchocelides,[581] glands.

[The larynx] varies in size according to age. It is augmented at the onset of puberty when the voice changes. Galen compared the size and width to the depth of the voice in Arte Medicinae 67, and Epidemics VII, 25. A sharp voice [results] when little air is moved swiftly from a narrow place.

It is composed of 3 cartilages, 3 common and 5 special pairs of muscles, glandular membrane, and vessels; veins, arteries, nerves. In particular it is composed of glottis, internal little tongue, cartilage and muscles.

[There are] three uses for the sternum: [1] rampart for 67v the heart and vitals; [2] binding for the ribs; [3] support for the membranes of the mediastinum.

Sometimes it protrudes outward, *out chested*, origin of gibbosity.[582]

Cartilage, mucronate, epiglottal;[583] xiphoid process;

[580] The prominence of the thyroid cartilage.
[581] A bronchocele is a swelling in the thyroid gland, but it is difficult to relate this with the division of the trachea.
[582] A distortion of the chest and spine usually applied to such conditions as scoliosis or kyphosis. In these conditions the sternum is frequently abnormal, but this follows the spinal deformity and not vice versa.
[583] The xiphoid process of the sternum. The term epiglottal may be used by Harvey in this sense instead of its more usual meaning.

malum punicum [xiphoid process]. B. When depressed inwardly it harms the liver and the stomach and causes atrophy; severe trembling; sign of tabes; a cavity. The cavity having been retracted and WH tension, a sign of obstruction. NB from friction of that part and compression, nausea [arises].

Special containing parts of the chest. The sternum 68r which is extended in the middle of the ribs [is made up] of 6 or 7 bones; more in children and fewer occur in old age. One of these having been corrupted, Galen [writes that] he saw the motion of the heart in a boy so affected, VII, 13, [De] Anatomicis Administrandis.[584] The cartilage in the end [is] ensiform, and sometimes in the aged osseous; for protection of the mouth of the stomach; sometimes perforated in women; sometimes divided into two parts.

It is connected to the diaphragm by a double slip [of origin] *with a hollownes* in the middle *where* WH NB *I sawe a wound cured that blew out a light and yet* not penetrating. *Here gather wind* often *with great oppression* in the upper part of the clavicle; but of these matters more exactly [in the account of] the bones. Triangular muscle; small, slender, fleshy. In dogs [it arises] from the clavicle.

Cartilages join the ribs to the sternum, especially of the seven true [ribs];[585] the rest hardly reach the sternum. Cartilages in the aged degenerate into bone, viz., in the superior and true ribs, not the false; wherefore such often have difficulty in breathing according to Hippocrates.[586]

On each side twelve ribs, seven true, the rest false, *short;* by degrees shorter as the body is bent at the sides; without compression lest it press the intestines [like the] *wheele of a coach;* wherefore in those with a longer body [there can be] thirteen or more, as [there are] 18 in a horse; in ser-

---

[584] *On Anatomical Procedures*, tr. Singer (London, 1956), pp. 192 f.

[585] Harvey makes the point that the upper seven ribs on each side articulate with the sternum by their costal cartilages. The eighth, ninth, and tenth ribs have cartilages that fuse with that of the seventh. In modern terminology only the eleventh and twelfth ribs are called "false" in the sense that their small cartilages have no connection with the sternum or with the other ribs. Cf. Harvey's statement below: "twelve ribs, seven true, the rest false."

[586] *Aphorisms*, II, 31.

pents as many as the days of the month; *grey hound eight Ribb:* for sign of long body in men &c.; as many as the months in a year; [Bauhin says] sometimes thirteen; [587] [according to] Colombo 11; [588] ribs and nerves &c., without sternum in some fish.

The highest [ribs are] the shortest. Ribs [are] very frag- ile in children, the obese [and those of] humid complexion. In [the case of] strong ribs the strength depends upon muscle. A sinus in the lower part [of the rib runs] lengthwise for vessels: vein, artery, nerves; wherefore very favorable [arrangement] in opening an empyema.

Subclavian muscles [run] between the clavicle and first rib, fleshy with oblique fibers. Intercostal muscles between ribs: exterior, interior, superior, inferior. Between cartilages [too], but only internal. Wherefore [there is] a controversy regarding number; some [declare] 68, [589] [but in fact] 44, external [and internal], viz., ii ii

$$ii \ ii$$
$$7 \ 7$$
$$5 \ 5$$

these between the bones of a brest-mutton. WH Perhaps they do not draw, as it does not appear to the senses; sustained in tonic motion *as yf the Ribbs wheare tyed with strings;* nevertheless it is argued that the external cause inspiration

[587] Bauhin, p. 356.

[588] Colombo, p. 485: "The ribs likewise twenty-two, twenty-five and twenty-six."

[589] The number 68 is explained by Crooke, p. 351: "From the undue distinction of the Fibres, some [e.g., Avicenna] have made 68. intercostall Muscles, when indeede they are but 44. reckoning those differing Muscles which are betwixt the bony parts of the ribs from those that are betweene their gristly parts. And whereas the distances of the ribs are eleven, six of the true ribs and five of the bastard ribs; they think, that there are foure in every distance of the two ribs, two where they are bony, and two where they are gristly." Harvey, after giving the number 44 as the total of the internal and external intercostal muscles, presents a numerical diagram to explain how this is arrived at. The appearance of these numbers in the published transcription is misleading, since there is no indication that there are really two sets of figures. In reality, Harvey is giving two sets of figures, indicating that there are two muscles, internal and external, on each side of the rib, and seven true and five "false" ribs. Thus with the twelve ribs there are eleven spaces and consequently twenty-two intercostal muscles on each side.

and the internal contraction. Some [maintain] the contrary, that the internal cause inspiration, some that both cause expiration.[590] WH I [maintain] that both [act] for both [purposes]; when both upward or downward at the same time all the ribs are moved. But these things [will be considered more] precisely [in the discussion] of the muscles.

The pleural membrane surrounding. Where pleurisy Δ tumors and phlogoses;[591] oedema and erysipelas. Sometimes thin at the peritoneum; investing the whole thick capacity; and the coverings for all arise from the bones of the back. It is proportionate to the peritoneum in substance and use.

Thin, strong, sinewy *the inner scin of a brest of Mutton. pellucid.* lightly lubricated with a watery humor; sometimes with profuse fat; aposteme Robert de Vill.[592]

Vesalius [says] that the mediastinum or intercepting 69r membranes arise from the pleura;[593] but [they are] thinner than the pleura; they divide the chest into two regions.

Sometimes, as in the case of the omentum, [the mediastinum is] filled with fat, but only in the exterior part, not in the cavities. The cavities [are] filled with fibrous connections [and] contain the heart suspended in the pericardium.

The ascending vena [cava] and the nerve of the diaphragm[594] [run through the mediastinum], wherefore the vena cava does not perforate the pleura as neither the spermatic vessels [perforate] the peritoneum. The esophagus also adheres with the nerves of the stomach and divides the chest into two regions like *two flagons,* one perforated for the sake of security. On this membrane lies empyematous pus since it rests with difficulty on the opposite side.

---

[590] Until modern times there has been confusion over the part played by the intercostal and other muscles in respiration. The modern view, briefly, is that in quiet inspiration the intercostal muscles raise the ribs and thus increase the antero-posterior diameter of the chest; at the same time, descent of the diaphragm occurs. In quiet expiration the diaphragm ascends and the elastic recoil of the ribs returns them to their resting position. In forced inspiration and expiration many other muscles are used.

[591] Phlogosis is an inflammation or flushing.

[592] Unidentified.

[593] *Fabrica*, p. 495.

[594] The phrenic nerve.

Thimus sweete bread Nutt of veale. The soft glandular body [lies] in the division of the vessels, which in this place are very confused; wherefore dissectors almost always err here; *heare they sticke the piggg.*

In the embryo [the thymus is] large, divided, like lungs; milky humor. Twice in two dogs, in one examined by vivisection [it was] divided, small, very white. In the other, [which was] dead, [it was] of ashen and reddish color; much larger whence a l'envoy. Wherefore small in a young, very large beef it appears as if not fully mature, and similarly in others.

Diaphragm, diaphratim [meaning] to distinguish; a sep- 69v tum because it separates transversely; from its disposition [called] phrenes by the ancients, because when inflamed it produces frenzy, or because as if the mind [were] there. Metonymy of containing in place of contained. *Midrefe: as sheer-reva, his office serving to both belly, he is stickler betwene them.* Therefore site between two venters obliquely, wherefore [according to] Fallopius *helpeth excretion*[595] *tied along the extremity of ye Ribbs not att ye top but* a little within because for the sake of safeguarding; wherefore *two budg-*

Common to vertebrae
Very close [according to] A[cqua]p[en- dente][596]

ets on each side. In a porpos longish as a sayle conteyning stom-ach liver and all things because *endeth in a poynt beneth.*

It joins the obliquely ascending muscles of the abdomen[597] under the extremity of the ribs *and with the trans-verse the vndermost.* It is connected at the rear *to ye two ver-tebrae*[598] where *it sendeth down by the ridg bone to slippetts of flesh,* processes [like] fingers. *Above* [it is joined to] the capsule of the heart and the mediastinum. [It extends] beyond the liver [and] stomach as muscle, with sinew in the

[595] *Institutiones Anatomicae,* in *Opera Omnia* (1606), I, 14: "between which and the abdominal muscles, the intestines are compressed, and from that compression what is contained in the intestines is violently forced out."

[596] Hieronymus Fabricius, *De Respiratione et ejus Instrumenta,* ch. 8, in *Opera Omnia Anatomica & Physiologica,* (Leipzig, 1687), p. 172.

[597] These are the internal oblique muscles of the abdomen. Actually the diaphragm is most closely related here with the transversus abdominis, for the two muscles arise together by slips from the inner surfaces of the costal cartilages of the lower six ribs.

[598] The crura of the diaphragm, as Harvey points out, are attached by "to slippetts" of muscle to the sides of the bodies of the upper two or three lumbar vertebrae. His description of the diaphragm is very good.

middle [to which run] fibers *as spiders webb two coates one of peritoneum the other of pleura.*

Nerves [come] from lower cervical vertebrae,[599] descending downward midway between sternum and vertebrae. [According to] Colombo *Inserted* above the pericardium on each side by three twigs.[600]

Thus because [it is] formed of muscle and [has] the 7or greatest sensitivity, its nerves *far fetich carefully and from nerer the fontayne most fayer and bigg.* 1, many passions inflame the phrenes; 2, convulsive movements in hysterics; also in children; *gerger laboring*[601] so much do they stretch out as they twitch that [there is no room] provided for breathing, so that suffocated they fall down; also in certain ones [the voice is higher pitched]; [cf.] man that counterfeited those hysterics; 3, likewise [it is] the seat of laughter; by fear of that agitation. 4, likewise *sighing*[602] *and sobbing this place sore.* Indeed sometimes inflamed [in laughter]. **WH** *and this rather then laughter* because laughter by breathing out, *and* by laughter *the belly ake yett;* by tickling around this part laughter.

It is
wrenched
*by
childrens
sobbing*

[According to] Aristotle because the sense is affected by the motion [of tickling];[603] but motion by the sensitivity of this [part]. Wherefore those brought to death in battle laughing; [according to] Aristotle more likely true than [the account of] a head cut off which spoke.[604] **WH** I have known of hysterical passion with much laughter. 5, neighing of horses from motion which agitates [the diaphragm]; 6, likewise by its motion hiccup [occurs]; the cause is pos-

---

[599] The phrenic nerve contains fibers from the fourth cervical nerve but also gets some from the third and fifth.

[600] Colombo, p. 256: "Among those nerves there are two which entering from below between the fourth and fifth cervical vertebrae, are inserted above the pericardium."

[601] I.e., hysterical breathing.

[602] The published transcription erroneously gives "*sighting*" rather than "*sighing.*"

[603] *P.A.*, 673a:8: "For to be tickled is to be set in laughter, the laughter being produced by such a motion as mentioned of the region of the armpit."

[604] *P.A.*, 673a:10: "It is said also that when men in battle are wounded anywhere near the midriff, they are seen to laugh, owing to the heat produced by the wound. This may possibly be the case. At any rate it is a statement made by much more credible persons than those who tell the story of the human head, how it speaks after it is cut off."

sibly in the gullet at the mouth of the stomach because [hic-
cup] occurs from inspiration; and by tension it is ended; or:
puppet bellows or bellows broken Hickop, wherefore also
from the liver, an affection of hiccups; and in jaundice. Δ
Sir William Rigdon all the stomach [with] yellow bile,
wherefore a little before death much hiccuping;[605] *ded mans
cough*, viz., by convulsion of [the diaphragm]; all those passions [are caused] by its exquisite sensitivity; it extends as
far as muscle, and action [results] from its movement at
the vertebrae &c. &c. &c.

Shape [When] *taken oute it is like a flare*. In the middle [is] con- 70v
vex into a cone. Like a sail extended to all the praecordia;
differs from all in site and shape.

[There is] a controversy among authors regarding its
origin and termination. This is goat's wool [i.e., of no sig-
nificance]. **WH** Origin where the fleshy extremity [ex-
ists]; the sinewy [portion], which occurs in the middle [is
the termination]. Great controversy over its action and
uses. No purpose served in referring to and arousing opinions and beliefs. **WH** I shall say what I believe, that in all
muscles there is action to contract themselves along the
course of their fibers; wherefore from convex it becomes
flat. On the other hand, some [assert that it contracts]
more into a cone. For it is necessary either that the extremities approach the middle as Vesalius [believes][606] or
the middle [approaches] the extremity and so downward
[movement results]; or in each way according to which
[is] more mobile, the extremity or the middle; [according
to] Vesalius when the extremities of the ribs have been adducted inward and upward, they are dilated slightly to the
sides and thus with the thorax inspiring, it is argued that
the chest is raised; **WH** as in fish. Gills or *larke-netts*. **WH**
But it is raised only with long inspirations[607] when it is elevated by the muscles of the spatulae. [According to] Colombo it is said that the chest is constricted and expiration

[605] Cf. n. 170.
[606] *Fabrica*, p. 495.
[607] Harvey is referring to the costal cage. In "long" inspirations other
muscles assist the intercostals to lift the ribs, including, as he says, "the
muscles of the spatulae [scapulae]."

168

occurs on the same basis as the motion of the extremities. Both are referred to vivisection; but X they suppose both more fixed at the middle than the extremities, as Δ WH with the ilia retracted; wherefore X [according to Du] Laurens by the action to form a cone; [608] just as there may be *retraction, contraction and action X for it can make a cone of itself.

*In the dead by laxity of the fibers

WH By contraction from a cone a plane is made and it is moved downward by impulsion at the middle; for the middle is more mobile than the edges; wherefore the thorax having been perforated in dissection and air entering, it can be drawn downward; and a cone having been formed by flatus &c., it is more difficult to inspire; and the thorax is dilatated by inspiration.

7r

WH On each side it elevates the ribs and depresses the intestines

But WH in inspiration [1] by a long sigh it is retracted; [2] in a quiet condition, the contrary.

Use, that it be [1] the septum of the venter: [dividing] noble [from] ignoble [parts]; vital [from] natural; lower [from] upper; [2] like muscles, by its motion. [According to] Aristotle the septum exists in all sanguineous and X feathered animals. [609]

Use

1, it protects the place and chamber of the heart from fuliginous, foul and crude vapors steaming from the kitchen, arisen from coction and from excrements.

2, it protects from crushing during distention of the venter of the uterus and colon by the flatus &c., of food; and in fermentation of the humors beneath it is inflamed, and [according to] Hippocrates such die. [610] A special use wherefore in fowl a *Gisard*, or where the stomach is cartilaginous and where there is no colon; on the other hand, where more distention occurs [it is] more robust as in the *Indy goate*.

3, in man *as an Apron* to support the heart and lungs in erect stature.

[608] Du Laurens, p. 343.
[609] *H.A.*, 506a: 4: "Further, all blooded animals have a heart and a diaphragm or midriff; but in small animals the existence of the latter organ is not so obvious owing to its delicacy and minute size." *P. A.*, 672b:11: " . . . the midriff . . . is called Phrenes in sanguineous animals, all of which have a midriff."
[610] *Prognostics*, cap. 5; Littré, II, 123.

1, inspiration by a secondary propulsion of the intestines and praecordia; Δ by agitation of respiring flanks. For song birds sing without [the use of] the diaphragm; where there are muscles of the abdomen as antagonists it corresponds proportionately; wherefore where they lack muscles it is lacking.

2, [to assist] peristaltic motion lower down; just as by [the pressure of] a hand [it assists] the excretion of urine, faeces, birth and flatus; for flatus having been retained, all *shite and groan* NB.

3, in its motion up and down [to fan] the praecordia, the refuse of the body and the gate of the liver in which many uses occur; wherefore [according to] Hippocrates [it is] the bellows of the [lower] venter; wherefore Δ 2, *a dogg* in quivering movement *lolling*[611] *the tong.*

4, NB WH perhaps it drives blood and spirits to the intestines; to the members.

Lungs PNEUMONES, [derived] from PNEUO, I 71v breathe. The *lungs* [are called] *lights* because spongy [and] *light;* the winnowers of the heart, repository of the spirit. *Then fill all this two regions* [of the chest]; thus [they are] *blown vp in a live creature.*

WH I saw them in the recently dead Δ, and before the *halter* had been loosened from those hanged; in the dead they settle downward and *soe* therefore they collapse, as [according to] Galen when something [pierces][612] internally as in wounds penetrating into the contained air.

There is a conjunction with the heart because the heart, as the source of heat, requires cooling;[613] [the lungs are] *fastned* in the rear to the vertebrae *of* firmness;[614] on each side [they are] *Cirrounded with the Ribbs and defended;* above [the lungs] *depend* from the windpipe, below [they rest on] the diaphragm, to the front [they are in contact

[611] Given incorrectly in the published transcription as *lelling.*
[612] Galen, *De Locis Affectis,* I, i; Kühn, VIII, 5.
[613] Aristotle, *P.A.,* 669a:5: "But animals that breathe are cooled by air. These therefore are all provided with a lung." This fallacy—that one of the functions of respiration was to cool the heart—lingered on until the eighteenth century.
[614] The lungs are not fastened to the thoracic vertebrae. As Harvey puts it so well, they *"depend* from the windpipe."

with] the sternum. [They] *then Encompass the hart* and are connected to its capsule. They are [also] connected to the heart through the pulmonary artery and pulmonary vein. Nevertheless in some animals [they are] not around the heart but [at] the lower venter, as [in] *frogges* [and in the] tortoise; [they are] far from the heart in serpents.

In man [there are] often fibrous connections to the pleura. **WH** I have often seen them suppurated where it is joined, and this more in the upper part. Therefore in such [cases] pain in the scapula and in the arm is explicable *as being less exercised*, because [the pleura] is connected only in men, and *thos* of sedentary life, for this connection is impeded by motion; therefore in the lungs of active persons, phthisis is unusual, as *hunters, singeing men* &c.

Therefore Caspar Bauhin, *a rare industrious man*, **X** believes that connections of this sort are to permit the lung to follow better the dilatation of the chest; and the chest has been hollowed for security.[615] [It is] a basic fact of Galen that the lung is moved when it follows the dilatation of the chest.[616]

The heart is situated at the 4th and 5th ribs.[617] [it is] triply in the middle, and all dimensions [are taken] from it: above, below, to the front, to the rear; to the right, to the left. Therefore [it is] the principal part because [it is in] the principal place, as in the center of a circle, the middle of the necessary body. 72r

In man the point is slightly to the left [in order] to heat the left side and to compensate for its weakness.[618] In the dead [it is] more withdrawn from the artery [pulmonary veins?] than from the vein [pulmonary artery?]. In other animals [it is] exactly in the middle of the chest. In them

---

[615] Bauhin, p. 451.

[616] Galen, *De Usu Partium*, VI, x; VIII, ix.

[617] A considerable portion of the heart lies behind the sternum and the costal cartilages of the fourth and fifth ribs on each side; this must be what Harvey meant by his statement.

[618] *P.A.*, 666b:7: "In all animals but man the heart is placed in the centre of the pectoral region; but in man it inclines a little toward the left, so that it may counterbalance the chilliness of that side." In fact the apex of the left ventricle lies in the fifth left intercostal space about 4 inches from the midline.

the base [is] a little to the left: left of right; on the contrary, in the ambidextrous [it is] in the middle [according to] Nicolò Massa.[619]

It is contained in a capsule [called] the pericardium, sheath [or] theca [which] therefore [has] a shape like that of the heart. [The pericardium is] suspended from the mediastinum from which it originates, and [it is] thick; according to Aristotle [it is] the thickest of all membranes.[620] Sometimes [according to] Colombo it is lacking with consequent syncope;[621] it is lacking or very thin in the dog and the rabbit, and [there is] none in birds. The veins, arteries and nerve [are] very slender.

Some claim that the pericardium is formed of veins from the axillaries. Outside [there is] a hard covering of very strong fibers,[622] but inward it is smooth. It is a slight but even distance from the heart in order to give space for its motions. Nevertheless in one suffering from phthisis I WH have seen [the pericardium] compressed tightly around the heart.

WH Therefore necessary that it be moved by motion of the diaphragm

It is joined to the pleura[623] at the 6th and 7th ribs, and at the sternum; to the spine and diaphragm through the mediastinum to which it attaches firmly so that Aristotle[624] and Vesalius[625] were deceived, who say that it is filled with fat;

[619] Actually Massa wrote (f. 53ᵛ): "It is said that those having the wide upper part of the heart verging to the left employ their left hand as if it were the right, but those who have a heart directly in the middle which inclines neither to the right nor to the left, are ambidextrous."

[620] Harvey either misread Aristotle or recollected erroneously, since the latter wrote (H.A., 519b:2): "The largest of all the membranes are the two that surround the brain, and of these two the one that lines the bony skull is stronger and thicker than the one that envelops the brain; next in order of magnitude comes the membrane that encloses the heart."

[621] Colombo, p. 489: "I dissected a student in the Roman Academy . . . who lacked a pericardium, and so repeatedly fell into syncope."

[622] The pericardial sac consists of two layers, the outer fibrous and the inner serous. The cavity between the heart and the serous pericardium is a potential space only.

[623] The mediastinal surfaces of the pericardial sac are covered by pleura, which separates the sac from the lungs, but, except in pathological conditions, the pleura is not fused with the pericardium. However, the pericardial sac is fused with the central tendon of the diaphragm as Harvey points out.

[624] H.A., 496a:5: "The heart . . . is provided with a fatty and thick membrane where it fastens on to the great vein and the aorta."

[625] Fabrica, p. 506: "The cavity [of the pericardium] is . . . completely lacking in fat. The exterior surface . . . also lacks fat, although Aristotle

172

for [there is] none either within or without, but in the mediastinum, for the two coverings [become] one at the mediastinum. It is perforated in five places by the pulmonary
vein, pulmonary artery, the [great] artery and the great
vein. WH X but it provides a covering[626] as [does] the
pleura [and the peritoneum in] the abdomen. Use: for the
strengthening of the heart as a rampart. [Du] Laurens
[writes that] it cools and humectifies,[627] perhaps because it
abounds in water.

The humor is like serum or urine, to prevent the heart 72v
becoming dry through continuous motion; also the capsule
[is found] in certain fish, as in the yellow sting ray, in
which the capsule contains a yellowish water [like] that
kind which [is found] in the thorax and abdomen. [Cf.]
generation, use and passion in certain animals.

Sometimes no [water is present], sometimes more in the
thorax. According to Colombo [it is found] in the living
dog as well as in the dead,[628] but in the dead more [com-
monly] in regard to those wasted away, as in persons
hanged in the sun and not dissected.

Piccolomini offers six opinions [regarding the source of
the pericardial humor];[629] all of no importance. WH I believe it is provided by nature lest the heart become dry;
therefore water rather than blood [issued] from Christ's

---

believed otherwise, asserting the pericardium to be fatty." Perhaps Harvey
was guilty of confusion of two authors he quotes frequently, since Colombo
wrote (p. 323): "But between the heart and the pericardium there exists
considerable fat which is clearly apparent and adheres to the heart even
though Aristotle and Galen assert that no fat exists around the heart."

[626] The fibrous pericardium blends with the external coats and the aorta,
superior vena cava, pulmonary artery, and pulmonary vein, but not with the
inferior vena cava, which perforates the diaphragm inside the pericardial sac.

[627] Du Laurens, p. 347: "Nor does this serous humor lack its use, which
is the final cause that the heart is always moist lest it be dried out by reason
of too great and constant heat."

[628] Colombo, p. 323: "You will find this humor not only in the living but
also in the dead. . . . I displayed it . . . often in public and private vivisec-
tions of dogs."

[629] Archangelo Piccolomini, *Anatomiae Prelectiones* (Rome, 1586), p. 204:
1, from the aqueous spermatic material; 2, from the fat of the heart; 3, from
the thick portion of the inspired air; 4, from the watery excrement of the
third concoction of the heart; 5, from saliva; 6, from liquid slipping in
through the trachea.

173

wounds.[630] In women and men of great [constitutional] humidity therefore [there may be] palpitation of the heart and suffocation from a too large quantity. [Cf.] Jasolino regarding suspicion of poison;[631] cachexia or poor quality. WH [Cf.] respecting remedies for poison.

Adam from the dust. I understand this as humor and heart. [True] water lacking a pungency and saltiness; WH such nitrous *slippery scowring as in Butchers hands.*

Thus [I have] *poynted out all the parts of this vpper belly only handling there situation.* Problems [and] observations regarding the rest; quantity, shape, use, services and worthy qualities. Only regarding its dominion WH. The heart in its running is always in motion.   73ʳ

1, WH [It is] the principal part of all, but not for its own sake, for it is more fibrous in its flesh, and harder and colder than the liver; but because of the supply of blood and spirits in the ventricles:[632] 1, therefore it is the source of all heat; 2, therefore in one recently dead the right auricle [was corrupted] because of an aposteme; 3, therefore in fish almost a lack of blood; and the larger it is the more spirituous and the hotter the blood. I believe that in so far as [the heart] is distended and not [wholly] contracted [it is] possible for life [to remain]; wherefore the auricles pulsate after the heart by the amount of blood it expels.

[630] The question of whether water or blood flowed from the side of Christ as it was pierced by a soldier's lance. Originally a question that had interested scholastic philosophers and theologians, it was taken up by anatomists relative to the question of whether or not water, as distinct from blood, might have issued if the pericardium had been pierced. The matter was treated *in extenso* by Berengario da Carpi, *Commentaria* (1521), ff. 336ᵛ–338ʳ, who decided that water as such could issue only miraculously. Later anatomists either treated the subject briefly, as Vesalius, or not at all.

[631] Giulio Jasolino, *De Aqua in Pericardio Quaestio Tertia* (Naples, 1576), no pagination [B³ᵛ]: "Some years ago, by order of the magistrate, we undertook the dissection of a certain noble whose sudden death had not been anticipated, and there was suspicion that he might have been poisoned. Although no indication of it appeared in the organs or any other sign of harm, we discovered the pericardium swollen into a huge mass which when lightly pressed with the hands, we observed its aqueous humor to flow abundantly from the aperture, and we decided that he had died suddenly because of palpitation of the heart resulting from this."

[632] Apparently Harvey was prepared, when he wrote this, to concede the ancient teaching of vital spirits mixing with the blood in the ventricles of the heart.

174

2, **WH** [The heart is] not first in origin, for I believe the ventricles (which in the foetus are both united as in fish) are made from a drop of blood which is in the ovum, and the heart with the other [organs], as in the awn of the wheat grain, all sprout at the same time from [a spot] of imperceptible size.[633]

Whether or not there is only a drop of blood in the auricles [at this time] whence communicating by heat into all the parts but receiving from none; like the citadel and home of heat, lar of the edifice, *fowntayn, conduit, hed*. Conical shape, like a pine cone; in *smelt* a pyramid; in birds a rounded cone; in the *mackerell* a tetrahedral, triangular shape; in the tortoise longish ; in some a double apex, as in the embryo.

In quantity the heart of man is large, wherefore timid, wherefore courage from a common intelligence; in length 6 fingers, in width 4. It is large in the timid hare, stag and ass, because the heat is less dispersed. For vigor results from the heat of blood; therefore [we have] the irascible [person]; on the contrary, [those] slow to fear, do not fear to become angry, for in them the bubbling of the blood [is] as a bubbling lake, but with great repose they swiftly cool down. Furthermore, the heart in its constitution is cold, small and thick, very compact and bold, for a small fire in a large space heats less than in a small. Therefore in those the disposition controls fear and anger, and the spirit follows the body. A hard heart [indicates blunt] sensibility, and a softer one, keen sensibility.

Scarcely any grave injury is suffered [by the heart] without death. _{73v}

1, **WH** I have scarcely ever seen injury [to the heart] in cadavers, nor is it consumed by phthisis in accordance with the belief of Galen,[634] but is [always] juicy flesh. **WH** On the contrary, the lungs always [are consumed in phthisis]

---

[633] In his *Anatomical Exercitations concerning the Generation of Living Creatures* (London, 1653), pp. 89 ff., Harvey gives a most poetical account of the development of the heart. His interest in embryology had been fostered by his old teacher at Padua, Hieronymus Fabricius.

[634] *Comm. V Galeni in Hippocratis Epidemiorum lib. VI;* Kühn, Vol. XVIII, pt. 2, pp. 203 ff.

whence Aristotle explains[635] old age through a dry lung, and tabes as a disease of old age, a wasting disease by the nature[636] of old age.[637]

2, The animal cannot die unless the heart is affected. **WH** Wherefore prognostication X. Yet it is rarely[638] observed. Colombo [describes] hard tumors,[639] but such kind perhaps a pulse or [due to] abscess in the left ventricle. Vesalius in book ij [mentions] glandular and blackish flesh in a man in whom [there was] a remarkably uneven pulse. Very often[640] a little stone in the heart in palpitation of the heart.[641] [Fabricius of] Acquapendente said a palpitating[642] heart. **WH** I [have seen] a very flaccid and pale [heart]. I heard that those wounded in the heart had lived for a time and had killed others. With the heart removed the *frogg* [can] *scipp*, the *eele crawle*, the *dogg* walks.

It is divided into KEPHALE, head, point, cone, cusp and apex. The base or root is wider because of the entrance of the vessels.

Division [of the heart] into parts. [1] External parts: pericardium, fat; crackling STĒAR according to Jasolino;[643] **WH** covering of the heart, tunic; coronary ar-

[635] The published transcription erroneously gives *deficit* instead of *definit*.
[636] The published transcription erroneously gives *nosos* instead of *na[tura]*.
[637] *De Respiratione*, 479a:10: "When owing to the lapse of time, the lung . . . gets dried up, [it] becomes hard and earthy and incapable of movement . . . the fire goes out from exhaustion." *De Longitudine et Brevitate Vitae*, 466a:24: "When [the materials constituting bodies] age, they become dry."
[638] The published transcription erroneously gives *caro* instead of *raro*.
[639] Colombo, p. 324: "the substance is very hard and it is dense and was made so that it would not be shattered even by strong movements."
[640] The published transcription erroneously gives a meaningless *scolsius* instead of *saepius*.
[641] This probably refers to an ante-mortem clot found not infrequently in the heart after death. These clots aroused great curiosity in the seventeenth century. In 1639 Edward May wrote a book entitled *A most certaine and true relation of a strange monster or serpent found in the left ventricle of the heart of John Pennant*, which contained a fantastic description of such a clot with references to other examples.
[642] The published transcription erroneously gives a meaningless *scabiosum*. Correctly it must be *saliosum*. The fact that Harvey remarks that "Acquapendente said" suggests the reference is to something heard directly from his Paduan teacher.
[643] Giulio Jasolino, *Quaestiones Anatomicae . . .* (Naples, 1573), f. 24ʳ, in reference to a body dissected in 1570: "when we liquefied the fat we discovered that it was liquefied with very great difficulty and afterward

tery[644] and vein; auricles or rather wings **WH**; ventricles; very slender nerves **X** Fallopius.[645] [2] Internal parts: containing parts, contained parts.

WH I do not know from what obstructed nerve death results, but Archangelo Piccolomini dreams.[646]

Parts containing the interior things: ingress of veins; ventricles and auricles; fibers, valves; septum; coronaries [vein and artery]; cartilage, bone; vessels of the embryo. Parts contained: spirit, blood; a white egg [is] contrary to nature.

---

it again congealed; but it did not preserve its hard substance, nor is this astonishing for reasons which will be adduced when, with God's favor, we shall speak of the water of the pericardium; and when the fat had begun to melt, in the liquefaction at first for a noticeable period of time we clearly heard it give off sound and crackle."

[644] There are, in fact, two coronary arteries.

[645] Fallopius, f. 209ʳ·ᵛ: " . . . I shall . . . describe to you the number and nature of the nerves of the heart. Beneath the base of the heart, where the pulmonary artery begins to bend to the left and where, in the embryo a notable arterial passage is present which unites the aforesaid vein to the aorta, there is a plexus or nervous network. . . . from which an abundant supply of nerve-like material surrounds the entire base of the heart and through which numerous small nerve shoots pass downward to be dispersed throughout its entire substance, although I am unable to follow them more minutely. . . . This nerve plexus takes origin from five shoots of the nerves, and although there are sometimes only four yet five are usually found. Of these, the first is that which I mentioned above and said arises from a large branch of the left of the sixth pair of nerves a little beyond the origin of the recurrent nerve, and carried down from there is reflected on to the left arterial nerve to ascend to the said nerve plexus. The second and third shoots arise also on the left side from the plexus which I called the plexus of the sixth pair when I described it in the neck. From this plexus on the left side, two nervules—although there is sometimes only one—arise which, descending to the base of the heart, are distributed through this plexus. There is also a fourth shoot on the left which others suggest arises from the recurrent nerve of that side, and this shoot, together with the second and third, descends and is distributed to the plexus. The fifth and last shoot, which is placed on the right side has a double origin. A nervule arises from the plexus of the right sixth pair which extends straight to the heart, and similarly another fairly large nervule emanates from the nerve of the same sixth pair which as soon as it has entered the thoracic cavity forms the right recurrent. The latter nervule unites with the former and produces a single and rather thick nerve. This nerve, which I have called the fifth, from its double origin runs hidden beneath the artery seeking the left clavicular region, and passes to the same network where it intermingles with the other four and becomes the single, dense, nervous origin from which many nervules are later carried as a stream to the heart.

[646] *Anatomiae Prelectiones* (Rome, 1586), p. 216: "This nervule is that one which sometimes, although rarely, is suddenly obstructed, whence sudden and unexpected death results."

[The heart] is continued by the root of the vessels, pul- 74r
monary artery &c. The pulmonary artery here [447] above
the aorta in [the form of] a cross.[448] Here the vena cava
enters, here the aorta [leaves].

WH Why the artery originates hence from the heart
rather than the vein I do not see. For the artery, like the
vein, sends a branch into the liver and the vein divides and
is continuous to the ventricles of the heart, and in some the
veins are not different from the auricles.

WH Query regarding the origin of the veins.[449] I believe
from the heart, the larger of which from the smaller, and
all from the center from which they grow and are extended
but the vein may be considered as a pathway; in the liver
where there is entrance of a small channel, and so the
principal gateways to the intestines.

But as a vessel and receptacle divided into many
branches, so the branches [are] like members of which the
principal one of all is the vena cava; of the cave or heart.
And here at the heart where is the very ample *fowntayne
hed*. In like manner many branches of nerves from the 6th
pair, of which the latter is the origin, but its origin is the
brain.

Furthermore, if as the containing and concocting vein,
where it contains and concocts more, there is the origin;
but that [is] in the heart because there is the most widespread and abundant heat for them. Then all the principal
[vessels] are not principal in respect to their origins, for
this [concerns] more the origin which is most principal.
Wherefore in regard to the vena cava and its branches,
that part of it is [most principal] which is the most principal place in the center of the body.

[447] The published transcription incorrectly gives *his* rather than *hic*.
[448] The pulmonary trunk arises from the right ventricle in front of the
root of the aorta. It runs backward to the left side of the ascending aorta and
then divides into right and left pulmonary arteries. The right pulmonary
artery runs under the arch of the aorta. The trunk with its branches is not
unlike a cross in appearance.
[449] Galenic physiology demanded that the veins have their origin in the
liver. Harvey doubts this and goes on to say, "I believe from the heart."
Later, in *De Motu Cordis*, he proves by clear-cut experiment that the blood
passes from the veins to the heart and thence to the arteries, thus disproving
the ancient theory of ebb and flow in the veins.

178

Vessels, vein and artery. **WH** But as in the pigeon the same covering of veins and arteries; wherefore doubtful whether a passage in them; and not in fish, except a fistula; wherefore the left ventricle is lacking, not the right as commonly [believed].

WH Perhaps in some, as I maintain, there is present a dispersible nutriment; lungs and left ventricle for dissipating heat.

[There are] two ventricles, right and left. **WH** I am 74v astonished at Aristotle since he describes three so precisely.[650] Not unless it is possible to exclude the left auricle from the ventricle, in which matter Galen rightly blames him since the ventricles are the same in the horse[651] and the sparrow.[652] Therefore **WH** I doubt whether animals have undergone such great alteration in the passage of time. [It is strange that] an author so diligent and faithful [should err].

Use of these two cisterns of blood and spirit. Right, semilunar circumscription in some animals invests the whole; in birds not reaching to the extremity of the point. **WH** Yet in the embryo as twin white nuclei, very wide, the flesh very lax and the wall very thin. [In some it appears to be] a kind of appendix to the left; see in the goose. More filled with blood, and therefore the hot right side, but the right auricle the last to pulsate. The left [is] exactly in the middle of the heart when the right has been removed; [it is] narrower with a triple wall [which is] thick; [I have seen] this in the embryo **WH**. Wherefore so very slender that it seemed lacking to Colombo.[653]

[650] *P.A.*, 666b:35: "... it is only in the largest hearts that there are three cavities. Of these three cavities it is the right that has the most abundant and the hottest blood. ... The left cavity has the least blood of all, and the coldest; while in the middle cavity the blood, as regards quantity and heat, is intermediate to the other two, being however of purer quality than either." Mondino described the heart with three ventricles, his account being borrowed from Avicenna. In the edition of his *Anothomia* printed in Strasburg in 1513 there is an illustration of a heart in which the "middle ventricle" is shown.
[651] The published transcription incorrectly gives *eque* rather than *equus*.
[652] *De Usu Partium*, VI, ix; Kühn, III, 442; Daremberg, I, 404.
[653] Colombo, p. 489: "I saw some in whom the left ventricle of the heart seemed to be lacking, it was of so slender size."

179

Here the spirit is perfected and is distributed hence into the whole body. X Galen De Usu Partium, VI, 9, VII. 5.[654] The right ventricle [exists] for the sake of the lungs; Δ of fish in which the right ventricle is lacking; because X WH.

There is no other organ of contained blood [so] filled to 75r capacity,[655] wherefore Aristotle, contrary to the physicians, [states that] the origin of the blood is in the heart, not in the liver, because there is no extravenate blood in the liver.[656] WH Blood is rather the origin of both, as I have seen.

Small bits of stonelike fat contrary to nature[657] are [sometimes] contained [in the heart], according to Bauhin.[658] WH Perhaps he was deceived by white blood that I have found in cachetics in the ventricle and auricle, and in all I have found little phthisic in the aorta; such white granules thrown off by a cough clearly ended in sanguineous clots; wherefore I believe too much concocted by the blood, for [they occur] in the hot and principal place; and of such sort that they float on the blood and are the washings of the fibers of the white blood; likewise dissolved by cold and moisture.

Skinlike auricles (black) for distention. The right [is] more wrinkled, more pointed, less hard and a little more fleshy and thicker than the left; it is composed of three kinds of fibers according to Dom. Leoni, p. 695.[659] Within I have seen it wrinkled by fibers, and like the ventricles so filled with blood as an abscess before it has hardened. It

---

[654] Kühn, III, 441–443, 525–527; Daremberg, I, 404–406, 465–467.

[655] The published transcription incorrectly gives *refertissimae* instead of *refertissimum*.

[656] *P.A.*, 667b:22.

[657] This description could apply to the plaques of atheroma which occur in the intima of the aorta.

[658] Bauhin, p. 412: "Often in the cavities of the dissected ventricles of the heart we have observed certain little bits, if not of fat at least very much like it, and it is greatly to be wondered at how it concretes there."

[659] Dominicus Leonus, *Ars Medendi* (Frankfurt, 1579), p. 695, gives a description of the auricles: "This wrinkled surface appears to the spectators as if covered with folds. However, the internal surface of the ventricles compares somewhat to the surface of the ventricles of the heart. . . . The auricles [are composed] of three kinds of fibers of this substance."

was put there especially for the high-spirited and irascible and for those overwhelmed by fevers.

Used like a storehouse for the heart; in fish a sanguineous bladder. In the embryo [there is] little blood in the ventricles, [but] much in the auricles. Three tricuspid valves, outward and inward, *like broad arrowes.*[660] Inward they are connected to ligaments at the point of the heart, outward to a membranous circle. These were first discovered by the amazing Hippocrates.[661]

WH X [according to] Galen the vena cava does not send a branch into [the auricle], but begins and is continuous thence;[662] wherefore the heart as the origin of the veins.

[It is] ridiculous to say that a branch is larger than the double trunk.

The fibers [are] wrinkled by lacertous ligaments.[663] In 75v some many, in others few and smaller. Observe that the largest [are] in those auricles in which the heart [is] larger; larger and deeper on the right than on the left; therefore Hippocrates locates the deepest chamber of the soul here.[664]

Use. **WH** To move the heart like a muscle when it is contracted; wherefore it is seen that systole is that contraction of the heart in which the ventricles are compressed by the fibers acting on all sides. In geese, ducks, *woodpeckers* &c., [there are] no fibers on the right just as no tricuspid valves; therefore the right ventricle does not pulsate in them with the very fibrous left.

The pulmonary artery because of its duty and covering in the lungs has a double trunk with a thicker covering lest the thinner part of the blood, which is the nourishment of the lungs, transpire. The different opinions of others are not offered.

[660] Refers to the valves in the right atrioventicular orifice. These valves, as Harvey says, are "connected to ligaments [chordae tendinae] at the point of the heart."

[661] *On the Heart,* ch. 10; Littré, IX, 87.

[662] *De Venarum Arteriarumque Dissectione,* ch. 1; Kühn, II, 768.

[663] In both ventricles the inner walls are made irregular by the presence of muscular columns called trabeculae carneae, which are more numerous in the left ventricle than the right. The walls of the atria are made irregular by the musculi pectinati, which are more numerous in the right than the left.

[664] *On the Nature of the Bones,* ch. 19; Littré, IX, 197.

The three semilunar valves[665] when tensed are gibbous but [when] relaxed [are] semilunar. The coronary vein goes about the whole base of the heart and sends branches to the point. [According to] Eustachius it is closed at its orifice by a semilunar valve.[666]

The pulmonary vein because of its duty and its covering [is divided] into two trunks so that [it presents] a kind of double orifice in the two lungs. Almost at the aorta they immediately divide; in the vessel of the child it resembles the portal in duty and origin.

The left auricle of the heart is more fleshy &c. Of this 2 tricuspids, *a bishops miter*;[667] from a membranous circle outward, inward.	[According to] J. Caesar Arantius[668] pulmonary artery with a capacity of ij fingers; pulmonary vein admits 4 fingers, wherefore its orifice is larger than that of the windpipe.

The aorta [is] near the entrance of the pulmonary vein.[669] 76r Between them [is] only one valve. The spirit [passes] hence into the great artery and thence into all. Here in man [is] cartilage, in the stag, bone; thereafter three valves, gibbous when tense, semilunar within and without when

[665] The pulmonary valves.

[666] The coronary sinus opens into the right atrium of the heart. Its orifice is guarded but not completely closed by a semicircular fold of the membrane lining the atrium. This valve was named after Adam Christian Thebesius who described it in his *De Circulo Sanguinis in Corde* (Leyden, 1708). It was well known to Eustachius, however, who clearly illustrated the valve and described it in his *Opuscula Anatomica* (Venice, 1564), p. 285: "The mouth of the coronary vein has a very delicate membrane in the form of a half-moon."

[667] A reference to Vesalius's assertion that the left atrioventricular valve resembles in appearance a bishop's mitre (*Fabrica*, p. 514).

[668] The published transcription erroneously gives "J. Caesalpinus Aretinus." Correctly the reference is to Arantius, or Aranzi, and his *De Humano Foetus* . . . (Venice, 1587), pp. 93–94; cf. F. N. L. Poynter, *J. Hist. Med.* (1957), 12:152.

[669] The pulmonary veins are placed behind the commencement of the aorta. It is the pulmonary artery that is nearest the aorta. Harvey then says that "between them [is] only one valve." His meaning here is not clear, for neither system is connected to the other at this point. The ductus arteriosus connects the pulmonary artery with the aorta in the foetus, as Harvey mentions below.

relaxed, so that now ij valves. 2nd, the iij remaining individual [valves are] for the pulmonary vein; here the coronary artery is joined with the coronary vein.[670]

The account of the transit of the blood and how the spirits are made.

Some believe that blood crosses through the septum[671] [through] interstices in the wall, gibbous on the right convex on the left, and accordingly they say it is porous **X**. However, [cf.] Bauhin's observation of the cooked heart of the ox.[672] Colombo saw cartilage.[673] In some, such as the tortoise, either no septum or a very slight one.

Vessels in the foetus; here a branch of the [pulmonary] vein as if a second [vein]; here the aorta degenerates into ligament. In the foetus [it is] as if a second aorta with a valve **WH**, and as if a third branch joined the pulmonary vein. Insertion.

Foramen ovale here[674] [is] an indication that there was a

---

[670] Just behind and to the left of the pulmonary trunk the anterior interventricular branch of the left coronary artery meets the great cardiac vein. This is probably the relationship to which Harvey refers.

[671] A reference to the teaching of Galen that blood passes from right to left ventricle through pores in the interventricular septum.

[672] Bauhin, p. 422: "Hence it is apparent that the blood is carried through those pores [of the septum] because nature never attempts anything rashly or in vain: but the pits are in the septum and terminate in very deep and narrow sinuses. This according to Galen. These little windings (spiracula) are very conspicuous in the heart of the ox after it has been long cooked."

[673] Colombo, p. 489: "In some I saw the septum by which the ventricles of the heart are separated, [to be] cartilaginous."

[674] Harvey's brief statement is elaborated in his *De Motu Cordis*, p. 48: "The right ventricle receives blood from its auricle and then drives it forward through the artery-like vein and its offshoot (the so-called artery-like channel) into the great artery. The left ventricle, in like manner, simultaneously receives blood (that has been directed from the vena cava, by a different route, through the oval opening) by means of the auricular movement, and by its tension and constriction it drives this blood through the root of the aorta into the same great artery.

"Thus in the embryo, while the lungs are idle and devoid of activity or movement, as though they did not exist, Nature uses the two ventricles of the heart as one for the transmission of the blood. And the condition of the embryo that has lungs, but is not as yet making use of them, is similar to that of the animal that has no lungs at all.

"The truth is thus as manifest in the foetus (as it is in the adult animal that has no lungs), namely, that the heart by its beat transfers blood from the vena cava and discharges it into the great artery. This it does by routes as free and open as would exist in man if the intervening septum were removed and the cavities of the two ventricles communicated with one another." As

valve WH; open in the embryo wherefore both auricles [are connected]; one void, as also in completed animals, fish, quadrupeds, oviparous. But of these and the use of them more exactly when the foetus [is described].

Galen gave[675] the best explanation;[676] the first modern writers passed by without giving consideration. Both remain open after weaning,[677] and also always imperfect in the calf, pigeon and goose. Furthermore in *Ratts*, &c.; other passages by which the auricles are mutually pervious; also in the human embryo as if a coronary vein. *Ratts* also [have] a vein at the right side in place of the intercostal which [is] below in man; here I believe at least upward.

A rod having been thrust into the vena cava [can pass] to the groin.[678]

The substance of the heart [is] dense, thick, hard flesh compacted like the kidneys; the blood of purplish color; in shrimps [is] white. WH Not equally blood-colored in all sanguineous beings, as not in cachetics in whom [its substance is] looser. The temperament of the heart [is] very hot since [it is] very full of blood. WH I have doubt of its dryness by reason of the body; flesh producing nerves is very humid by reason of the contained blood. In those in

76v

Singer points out in his *Evolution of Anatomy*, the fossa ovalis was illustrated by Eustachius, who claimed it as his discovery. Singer also states that this had been described by Sylvius and is shown in a figure of Vesalius. Vesalius' figures do not show the presence of the fossa ovalis. His figure of the interior of the right atrium shows only the openings of the venae cavae and the coronary sinus.

[675] The published transcription erroneously gives *explicabit* instead of *explicavit*.

[676] *De Usu Partium*, XV, vi; Kühn, IV, 243–244; Daremberg, II, 148: "One must admire nature which, at the time when the lung had need only of developing itself, furnished it with a pure blood, and when this lung became capable of moving itself, gave it a flesh as light as a feather so that it might be easily dilated and contracted by the thorax. It is for this reason that there is in the foetus an opening serving for communication between the vena cava and the pulmonary vein. It is of such sort that this vessel serving from vein to organ made it necessary, I think, that the other [pulmonary artery] fulfill the service of an artery; it is for this reason that nature produced it from the aorta."

[677] The published transcription erroneously gives *ablectationem* instead of *ablactationem*.

[678] This experiment is used by Harvey to demonstrate that the superior vena cava, the cavity of the right atrium, and the inferior vena cava are in one vertical line.

whom the ventricles [are] small, the veins [are] small, but I believe rather that [the size of the vessels] depends on the auricles. For on the other hand, as I said before, the heart of the timid [is] large.

Temperament of the heart [can be judged] from the pulse, breathing, hairs of the chest and axilla; from habits relative to the strength or weakness of the body.

The motion [of the heart] is commonly believed [to run] 77r from the point to the base; when [the point] has been drawn [toward the base], dilatation of the ventricle occurs and there is diastole. On the other hand, when relaxed, systole takes place at which time the chest undergoes diastole. When the heart has been dilatated in diastole the artery is dilatated, and that particular motion of the heart is diastole. Systole is concerned with death, and after diastole occurs the internal rest of the heart and pulses.

Having observed [the motion of the heart] for whole hours at a time, I was unable to discern [these things] easily by sight or touch, wherefore I propose that you ought to observe and note. It seems to me that what[679] is called diastole is rather contraction of the heart and therefore badly defined, or it is X as they say; or at least diastole [is] distention of the fleshiness of the heart and compression of the ventricles.

[According to] Colombo, p. 474,[680] when the heart is dilatated the arteries are constricted and again, upon constriction of the heart the vein is dilatated. NB When the heart is carried upward and seems to swell, then it is constricted; but when it is stretched out as if relaxing, it verges downward, and at that time the heart is said to rest; and then there is systole of the heart, because it receives more easily and with less labor; but when it transmits, more strength is required; nor may you consider this of no importance even though you will find not a few who believe with certainty that at that time the heart is dilatated, when in fact it is constricted.

[679] The published transcription erroneously gives *aut quum* instead of *autem quod*.

[680] A comparison of the pagination of the various editions indicates that Harvey made use of Colombo's book in the edition of 1590.

185

See how arduous and difficult [it is] to discern, either by 77v
sight or by touch, dilatation or constriction and what is
systole and what diastole.

Extension is its proper motion for it is animated; by relaxation[681] it is enervated. Extension is systole; 1st, it
strikes the chest by extension; 2nd, from soft it becomes
hard so that except when it is extended it does not seem or
appear larger; 3rd, the auricles contract sensibly and give
forth a lighter colored blood; 4th, at the same time by
touch the pulse of the artery is felt, as if the vena cava were
jerked.

The pulse begins at the auricles and progresses to the
point, wherefore as if [there were] two wings **WH**. Nevertheless the heart beats when separated from the auricles
and the auricles awaken the somnolent heart. It must be an-
swered that the heart by that first breath of air, afterward
by degrees with a delay interposed; finally not completely.[682]

In fish it is clearly compressed by extension and blood is
given forth. Also parts of the heart are dilatated according
to the flesh wherefore it is necessary that it be constricted
by the ventricles. Since the heart responds slowly it twists
its point a little bit to the left. Motion and life gradually
abandon the point wherefore [life remains longest] in the
final ventricle, and the final [ventricle] is the right. [There
is] internal rest after extension. If the heart is perforated
during extension blood flows and spurts forth wherefore
[that is the time] of systole, and at the same time the pulse
of the arteries [is perceived] by touch; [blood] leaps forward from the arteries and from the pulmonary artery.
After an hour the pigeon: with the finger the right auricle
will pulsate.[683]

Query whether the pulmonary vein pulsates **WH**. 78r
Whether the pulmonary artery. Query: with the pulsating
heart cut off, whether the ventricle can open and constrict.

[681] The published transcription incorrectly gives *relaxatur* instead of *re-
laxatione*.
[682] The views expressed in this paragraph are amplified and far better expressed in *De Motu Cordis*, pp. 33-34, 39-40.
[683] This experiment is quoted in full in *De Motu Cordis*, pp. 35-36.

[Cf.] Vesalius and Colombo.[684] When in a vivisection the lungs are inflated in varying degrees, the pulse is altered; when they collapse there occurs a creeping surge. Vesalius observed nothing more beautifully.[685]

Hence an error held now for 2000 years, for which rea- 78v son I have given attention to it, because so ancient and accepted by such great men.

1st, they speak of diastole and systole, and for diastole only dilatation in the fleshiness of the heart and constriction of the ventricles. 2nd, [they consider] extension, whether of systole or diastole, as special motions of the heart opposed to relaxation. 3rd, by extension [the heart] puts forth blood and produces a pulse, for Erasistratus [argues] against Galen['s idea] *as in a glove*,[686] and the argument of Galen from a hollow reed[687] [is] impossible.

[684] *Fabrica*, p. 570: "In the conduct of this sort [of vivisection] it is instructive to grasp the base of the heart and quickly with a single ligature to cut off the vessels extending from [the pulmonary vein]; then excise the heart below the ligature, loosen the bonds by which the animal has been tied down and permit it to run about." "I am accustomed to open the right ventricle of the heart and to observe here that nevertheless the heart still moves; then when the left ventricle has been opened, or the middle portion of the heart divided transversely and removed, I advise the spectators that they may observe the heart still moving and not yet at rest." Colombo, p. 481: "to behold a very astonishing thing, with a knife quickly open the thorax [of a dog] and grasp the heart; then allow your skilled assistant to ligate the heart swiftly with a curved needle and thread, and loosening the bonds by which the four legs of the dog had been secured, set it on its feet. And there is nothing more astonishing than to behold a dog without a heart barking and walking. "

[685] *Fabrica*, p. 572: "The rhythm of the pulses of the heart and arteries having been observed for some time, the lung must again be inflated, and by this means, than which I have found nothing more pleasing in anatomy, much understanding of the differences of the pulses can be acquired. For when the lung has been collapsed and flaccid for a long time, an undulant surging and vermicular motion of the pulse of the heart and arteries is observed. With the lung inflated, it again becomes strong and rapid and displays remarkable variations."

[686] Erasistratus considered that the heart dilates and contracts by its own force while the arteries dilate or pulsate by reason of the pneuma forced into them by the heart's contraction (Galen, *De Hippocratis & Platonis Decretis*, bk. 6, ch. 7; Kühn, V, 563). In *De Motu Cordis* Harvey remarks: "An idea of this generalized pulsation of the arteries consequent upon expulsion of blood into them from the left ventricle can be given by blowing into a glove, producing simultaneous increase in volume in all its fingers."

[687] This brief statement of Galen's argument is expanded in *De Motu Cordis*, pp. 13–14, as follows: " . . . I think that I can readily and openly show, and have before this publicly shown, that arteries increase in volume because they fill up like bags or leather bottles, and are not filled up because

But on the contraction of the ventricles the arteries are dilatated, wherefore [the arteries] fibrous for contraction, for it is necessary that the blood enter into the ventricles by means of force and relaxation; whence they are imbued with an increased color.

Hence the pulse of the artery [is] not from an innate faculty of the arteries, as according to Galen 13,[688] but by the heart thrusting forth [as is indicated] by autopsy in the live and dead, by reason [and] by experiment with ligatures.

Hence [the blood] is drawn into the part from the ligatures and thence the reason why tumors [arise] from pain and inflammation and the attraction of the humors, viz., through the arteries. Hence with the arteries obstructed [one gets] lividness and mortification.

Then [one can argue] from the position of the valves. 79r For they impede propulsion into the heart where it is acted upon; then the arteries having been compressed the blood rushes forth [but] air does not enter; and so [the blood passes] into the ventricles of the heart.

---

they increase in volume like bellows. This, it is true, goes contrary to the experiment described by Galen in his book under the heading, 'That the arteries contain blood.' He exposes an artery, incises it along its length, and inserts a reed (or hollow pervious tube), thus ensuring that the blood cannot escape and closing the wound. 'So long,' he says, 'as matters stay thus, the whole of the artery will pulsate. So soon, however, as you pass a ligature and knot it over the artery and the tube, thus clamping the arterial coats to the reed, you will cease to observe arterial pulsation beyond the knot.' I have not done Galen's experiment and I do not think it could well be performed in a living subject, because the blood would erupt too forcibly from the arteries; the reed, too, without a ligature will not close the wound, and I have no doubt the blood will leap out through the reed and beyond it. By this experiment, however, Galen seems to prove that the pulsatile power spreads from the heart through the arterial coats, and that the arteries during their dilation are filled by that pulsatile force because they dilate like bellows, and do not dilate because they are filled like leather bottles. The opposite to this is, in fact, apparent both in arterial section and in woundings. For the blood escapes from the arteries in forcible spurts and, even if the distance it travels is now greater and now lesser, the spurt is always in the diastole of the artery and never in its systole. From which it is clear that it is the force of the blood which causes dilation of the artery."

[688] *De Facultatibus Naturalibus*, trans. Brock (London, 1928), p. 315: "all the arteries possess a power which derives from the heart, and by virtue of which they dilate and contract." The published transcription gives *valvularum*, which, however, can only sensibly be *arteriarum*.

4th, that in systole there occurs a longer internal rest, wherefore in some animals after a count of 12 pulsations, and that death [may occur] after systole or at least extension.

WH But that [the heart] by extension propels and emits [blood] is clear from these things.

[1] by the experiment of ligatures gangrene [results]; then from the position of the valves; [blood] leaps forth from wounded arteries; the artery [has] a stronger tunic. That which is perceptible especially in the auricles.

2, for the heart is paler in color in its vigorous extension, as in frogs, fish &c.

3, from a wound, the blood, at that time in the ventricles, arteries and pulmonary artery, spurts forth.

4, the heart responds to the auricles, so that what has been driven into itself is driven forward.

Rationally also, 1, the experiment of Galen with a hollow reed[689] is impossible.

2, if the arteries and ventricles [pulsate] simultaneously, therefore not from the heart; for the arteries simultaneously [pulsate] for the reason that the valves impede, and the heart at the same time subsides.

3, from the pulse of the arteries and of the pulmonary artery—not the pulmonary vein; and in them a stronger pulse; [the pulmonary vein] more fungous, and [has] great differences from veins; on the other hand in pigeons &c.

4, in lizards [note] the position when it is contracted and compressed, as **WH** in the auricles; this is for the sake of the pulse, wherefore in those in which stronger and greater pulse, more &c.; on the other hand, in those in which the lung more fungous none on the right; [cf.] goose, duck, *woodcock*.

5, because a forceful operation (for it enters easily in relaxation); but it is strengthened in extension, and from [being] soft [it] becomes hard and is extended and acts; on the contrary in relaxation it is weakened as in fish, Gabaris &c.

6, the fibers having been contracted, the walls are ex-

[689] Cf. n. 687.

tended lengthwise, to compress laterally like the muscles of the abdomen; wherefore beating and impelling; pulse of the blood, compulsion and an impulsion, *as in a glove.*[690]

From these things it appears **WH** that the action of the 79v heart is in accordance with what is moved; blood [is moved] from the vena cava into the lungs through the pulmonary artery, and from the lungs through the pulmonary vein into the aorta. With the heart relaxed that blood first enters into the right ventricle from the vena cava,[691] [and] into the left [atrium] from the pulmonary vein.

The heart having been extended and contracted, just as by a kind of force it propels from the right [ventricle] into the lungs, from the left into the aorta; wherefore [occurs] the pulse of the arteries and much speculation by Galen regarding the pulses of the heart, especially of the internal rest, in de u[su] p[artium][692] VI, 8, and systole, although I believe never realized by touch; Galen ch. 12 [De] A[natomicis Administrandis], bk. vij.[693]

Action   *thus relaxed receyves blood. Contracted propell it*[694] *over.* In the whole body of the artery compares *as my breth in a glove.*

But for the sake of what thing? [cf.] Aristotle. Of nothing but passion, as in boiling pottage. **WH** But [when the heart] is wounded, it emits not spirit but blood. **WH** But the structure of the fibers and valves indicates that for the sake of something. [What of] the artery. **WH** Whether a ventilator for the motion of the bubbling blood. Whether parts are warmed by the arterial blood, wherefore the arteries[695] having been obstructed a part is cooled. **WH** Whether it opposes the dissolution of heat in fevers and

[690] Cf. n. 686.

[691] It will be noted that here Harvey uses the term vena cava as synonymous with right atrium.

[692] The published transcription erroneously gives dep p."

[693] The published transcription erroneously gives "(12 A—o)." Although the manuscript is difficult to decipher at this point, the discussion suggests this notation must be a reference to De Anatomicis Administrandis, ch. 12 of bk. 7, in which Galen discusses vivisection of the heart.

[694] The published transcription erroneously gives "Puppet is" instead of Harvey's important *propell it.*

[695] The published transcription erroneously gives *articulis* instead of *arteriis.*

pathemata of the mind [arisen] from heat; in boys rarely.

WH Hence in animals the arteries have two very thick 8or
tunics, and especially in adults in which the pulsation of the
heart is stronger so that the artery sustains a [greater]
force [than the vein].

WH Hence what has impressed no one: that the pulmonary artery is thicker [because] it sustains the pulse of the
right ventricle in adults, and the artery [is thin] in the embryo. Hence neither the vena cava nor the pulmonary vein
[is] of such structure, because they do not pulsate but
rather [the blood] is drawn [from them]; and this because
the opposed valves break the pulse in the heart and in the
rest of the veins.

WH Wherefore there are many valves in the veins opposed to the heart; the arteries have none except at the exit
from the heart. Hence the first veins are pulsating, the latter are nonpulsating.

WH Whether [the heart] strikes the ribs at the apex,
whether at the sides, is in doubt because the ventricles are
at the fifth rib where it is felt. The tip at the sixth or
seventh [ribs][696] because the heart is contracted and extended in its middle.

[According to] Aristotle in all sanguineous [animals] the
heart is proportionate to the rest [of the parts];[697] WH X
in the shrimp in which the heart [is] open and white.

WH From the structure of the heart it is clear that the 8ov
blood is constantly carried through the lungs into the aorta
as by two clacks[698] *of a water bellows to rayse water.* By [ap-
plication of] a bandage [to the arm] it is clear that there is
a transit of the blood from arteries into veins, wherefore
the beat of the heart produces a perpetual circular motion
of the blood. Is this for the purpose of nourishment, or

[696] The apex beat, corresponding to the apex of the heart, is usually felt
and sometimes seen in the fifth left intercostal space about 4 inches from the
midline.

[697] *P.A.*, 666b:22: "In animals of great size the heart has three cavities;
in smaller animals it has two; and in all has at least one." *Ibid.*, 666b:35:
" . . . that it is only in the largest hearts that there are three cavities." But,
667a:20: "the heart is of large size in the hare, the deer, the mouse . . . in
pretty near all other animals that are . . . timorous."

[698] Cf. n. 541.

more for the conservation of the blood and the parts by the distribution of heat; and, in turn, is the blood cooled by warming the parts made warm [again] by the heart? Shape of the lungs; bulk, number, substance and color 81r of the lobe &c. [There are] various shapes in various animals; [in the] *weosel* [and] *stoote* [it is] *long*, [in the] *doog*, *deepe*; in man, *rownder and more bulkey*. Shape [results] from the containing place, wherefore nature is not solicitous of the shape because the shape has none or little relation to the action. Convex upward.[699] Below [is a] *hollow giving way to ye liver*; within the *hollow giving rome to the Hart they Embrace. Thus blowne vp like the hoofe of an ox:* 1st, in regard to the cavity; 2nd, *the slite;* 3rd, *the edg;* 4th, *the distance over the harte in the slite;* 5th, in the convexity &c.

The bulk is such that when [the lungs are] inflated they fill that whole cavity. Those in whom the bulk is greater the *longer the breath* [may be held], because *the* [lungs are] *more dilatable*. On the contrary, [when] obstructed and compressed, [the individual is] *shorte winded*. Wherefore when the smooth arteries have been obstructed they fill with crude vapors and humors, or compressed or consumed by putrefaction, [such are] *shorte* [of] *winde* and *rather pant then breath*.

Wherefore the bulk [of the lungs is] for the sake of longer inspiration. Therefore in those there is much blood and the heart [is] hot, the lungs are larger, especially in the bronchi; the sign of this is the width of chest and the amplitude of the nostrils. Therefore all such [are] energetic, and in those hot in regard to the heart the nares are amplified by the lungs; wherefore feathered creatures *play;* as in those hot-tempered; in runners; in horses; wherefore in many animals large lungs in which heat from increased use; but in others because they breathe rarely, cold; as frogs, serpents, tortoises &c.; the lungs [appear] as a cluster of vesicles[700] and that in the abdomen, which inflated greatly

[699] Because of the obliquity of the first rib on each side, the dome of the pleura, and therefore the extreme apex of each lung, extends upward into the root of the neck for 3–4 cm. above the first costal cartilage. The top of the dome of the pleura is then at the level of the neck of the first rib.

[700] The published transcription incorrectly gives *vehiculorum* instead of *vesiculorum*.

PAGE 80ᵛ OF THE NOTEBOOK, SHOWING HARVEY'S FAMOUS
STATEMENT ON THE CIRCULATION OF THE BLOOD

distends the lower venter; wherefore *swell like a toad*, as frogs in the summer when they breathe. They have a small caruncle, in quantity the particle of a nut. Since large lungs [are required] for them so that rarely breathing they may remain for a long time under water or earth, both because [they are] cold and because they yet require a lung of large capacity, nature [provided] lungs of this shape and as compressible as possible, which [have] the power of distending into a very large mass.

Number: [the lungs] are divided into right and left 81v through the middle, hedged about by membranes and separated from the pericardium. This [is] for the sake of security, if one [is] injured, obstructed or consumed. **WH** I believe *as breath att one nosetril*[701] *and chaw at one side* so often only one part of the lungs is used and that in turn. **WH** Δ *ye one part hath served sum a long time*, indeed, *one little peece of one side*; so that it is astonishing how little.

In those in which the viscera are bipartite this division is more exact. In dogs and in birds, in which clearly [there are] two lungs; in all ovipara in which the division in the lobes is also more apparent than in man. The number of lobes [is] usually 5 in all. Ancient men wrote that the fifth and part of the right were a cushion for the vena cava. Therefore [according to] Avicenna the left lung [is] smaller than the right, yielding to the heart;[702] wherefore the heart pulsates, being more on the left. Recent writers [criticize] Galen **X** because they find only 4 [lobes], and they say that he *exercised in apes* [rather] than in men.[703] **WH** [I] *suspend my censure* [because] as in the liver, perchance [there were] then more [lobes] as now rarely [occurs]; **WH** for yesterday I saw 5 in an embryo. Giulio Jaselino [found] 7;[704] some only 3, some only ij; some none; and it appears that

701 This peculiar idea is, of course, a fallacy.
702 *Canon*, Lib. III, fen x, tract 1, cap. 1.
703 Galen, *De Usu Partium*, VI, iv (Kühn, III, 421), speaks of two lobes on the left and three on the right, and at VII, i (Kühn, III, 517–518), speaks of four lobes, two on each side, but adds that there is a fifth lobe on the right side which serves as a cushion for the vena cava.
704 Giulio Jasolino, *Quaestiones Anatomicae* . . . (Naples, 1573), f. 23ᵛ: "[in 1570] in the second cadaver we noted that the lung was clearly divided into seven lobes."

4 [are] not necessary, and very probably Galen and Avicenna have been calumniated.

WH NB *the fewer divisions* the more exact and, on the other hand, the more coherent. Use of the divisions for the sake of security, one part having been corrupted since even without division it would be necessary to dilatate within itself, and because the *body bowed together* without compression.

Their substance [is] spongy flesh *like prowd flesh*; fungous. In some more fleshy, in others more fungous; in the young; in the old; and in drier temperaments more fungous. Hence diversity in regard to temperament, some hot some cold, but less than the liver. But hot as filled with blood and humid throughout; if parenchyma they are smooth, soft, rare, lax, spongy and as *froth*. All these denote humidity. Wherefore [according to] Galen the lung [is] very hot, soft and also very rare and light, in constant motion.[705]

[They are] filled with veins and vessels, of sanguineous 82r color from which the heat is collected. WH I believe that they are as much more humid than the liver as they are softer, and the hotter as they abound more in blood, and that in regard to the hotter arterial blood; and that no other part of the whole body so abounds in blood. [The veins and vessels] appear less hot in color in the liver and spleen because they are paler, and from pale become flavescent.

WH I have observed: 1st, in the first conformation as white as snow; 2nd, in the embryo before the drawing of breath of the same color as the liver in children before birth, *and in two whelpes the one borne ded;* wherefore [according to] Avicenna the air is responsible for the white color,[706] so that the colors are ex accidente; 3rd, in the sick, a *swarty purple blemish*, as in peripneumonia, filled with blood, *duskey ash color, a durty greye leadish* [color], and in apostemes with and without livid veins [they become] *more white and yellow cley color*, contracted. In hectic fevers

---

[705] *De Usu Partium*, VI, x; Daremberg, I, 411–412.
[706] *Canon*, Lib. III, fen xxi, tract. 1, cap. 2: "For the lung is not red except in the embryo since there it does not breathe; it is nourished by red, subtle blood and does not become white except by mixture with air."

in man as well as in my apes, *seacolored* without drink. Wherefore NB there may be scholastica, there may be hectic fever of the solid parts,[707] and fever of a part of the heart, and 3rd degree hectic fever or marasmus, the part having been seized by a radical humidity. I have never yet [seen] it so for the heart, but [in] the lungs very often.

WH As usually [in] all who die of tabes, the lungs [are] dried out, fungous and contracted, wherefore Aristotle aptly [remarked] that old age is a natural withering,[708] and tabes is a disease of old age.

WH Certainly the whole temperament very much follows the constitution of the lungs; wherefore in those in whom [the lungs are] fungous and contracted, so the signs of this are roughness of voice, sharp constriction of the chest, high shoulder joints, *quickly grow and looke owld;* and they readily waste away. Likewise in animals in which the signs of hot lungs are depth of voice, amplitude of chest, large windpipe, great inspiration and hotter expiration; as in those greatly excited the amplitude of the nostrils as in the horse; the fire of the poet flares from the nostrils. All those [are] signs of boldness because of hot lungs.

On the other hand, contrary signs [indicate] timidity 82v because of coldness of the lungs.

The division of the lungs into parts containing [and] 83r contained.

Aaer. Contained: blood, air as one recently dead, like vesicles. In the tortoise *like a heape of blathers* [and in the] *porpos froth like aer and water* [mixed]. Therefore in some more blood, in others more air. For I believe that the hotter animals [are] the more their lungs are filled with blood. Wherefore [according to] Aristotle man has the hottest blood as his lungs are the most completely filled; wherefore [man is] the one being which moves erect;[709] hence

---

[707] Paul of Aegineta, trans. Adams (London, 1846), II, 33: "The hectic fever is not only seated in the fluids and spirits, but also in the solid parts."

[708] *De Longitudine et Brevitate Vitae*, 466a:22, 466b:12.

[709] *De Respiratione*, xiii: "the higher animals have a greater proportion oı heat . . . those with the most blood and warmth in the lung are of greater size, and that animal in which the blood in the lung is purest and most plentiful is the most erect, namely man."

196

vivipara hotter than ovipara in which the lung is more fungous [and contains] less blood. Therefore in the case of the latter the size of the body is less, preternaturally contained in sicknesses.

Passion. Apostema, great and slight ulcers, *like hoggs measels*. Calculi from gypsum, hair; *like chalkestones*. A supply of ichorous matter wherefore asthma [results], as in ulcers of the legs treated with mercury; and podagra.[710] WH Δ; wherefore dropsy, lethal cough and asthma; for a sign that the matter already holds the citadel of the body.

Containing parts: nerves, vessels, of which first the covering; bronchi, branches of the windpipe; flesh, parenchyma.

No fat except diffused WH; wherefore the flavor of young beef, *lambs, pork;*[711] indistinct nerves from the 6th pair after the withdrawal of the recurrent; wherefore slight sensitivity; wherefore pain suddenly concealed with increased asthma. A sign that [there is] metastasis in peripneumonia; wherefore it putrefies because *eaten a peeney* without pain.

The vessels first called smooth arteries.[712] Bronchi [ex- 83v tend] into the rough artery [trachea]. This [distinction] as seen between [the smooth and rough arteries] is to no purpose and can [now] be removed.

Flesh [is] spongy as before. The membrane [arises] from the pleura: *thin, smoth*, soft. Little foramina wherefore the transit for suppurated matter, but it is doubtful how it received the matter and yet detains air; and whether the lung *sup vp*; by distention or by contraction. WH I believe by expiration when the covering is relaxed, not by inspiration because it is stretched and the passages are

---

[710] The term applied in the seventeenth and eighteenth centuries to gout in the feet. Cullen later used the term to mean gout in any part of the body.

[711] The published transcription erroneously gives "port."

[712] It seems fitting that Harvey should use this phrase for which Aristotle was responsible. Regarding all blood vessels as the same, Aristotle used the term "artery" to describe the windpipe. In later times when a distinction was made between veins and arteries, the latter from their structure were called "smooth arteries," and the windpipe was called the "rough artery" or, in Latin transliteration from the Greek, "trachea arteria."

closed, wherefore the air does not go forth. **WH** But the works of nature are to be marveled at.

A principal part of Galen,[713] such as in no other part, as the crystalline humor of the eye; nothing more advantageous &c.; but 1st and principally therefore neither veins nor arteries &c. But it is in doubt whether the bronchi are parenchyma, for both have the quality of an enclosure.

WH But more of the bronchi.

WH Fleshy but parenchyma beyond, to concoct blood so that spirituous arterial blood may be made. For those in which the work [takes place have] hotter, thinner [blood]; *sprightly kind of aliment* necessary. Therefore **WH** Δ in all those in which arteries, fleshy lungs; and on the contrary, in those which [have] none or like fungous, veins or entirely without flesh, as frogs, tortoise. Thus in those without flesh and without arteries or at least arteries which do not differ from the veins.

Action of the lungs: [1] motion of dilatation and con- 84r striction; [2] action of alteration and of concoction: I say common or at least a common preparation. Therefore the 2nd actions [have] two principal parts, a second motion viz., of bronchial alteration [and] viz., parenchyma.

Of [the cause of] the motion there is much controversy: 1st, whether [the lungs] move themselves or are moved by another; 2nd, what is the mover, whether heart, thorax or muscles; 3rd, if [the lungs] move themselves, how and by what; 4th, if moved by another, [is it] either partly by another [or] partly by themselves.

Whether the motion is voluntary or involuntary. **WH** In their motion they do not collapse as in the dead but are retracted and are more dilated than now [in the dead] because the chest is dilated. Motion [comes] from two contrary motions whence the two which. . . . [714] [According WH Bladder in to] Galen and the school of physicians derived from him, bellows that they are moved by . . . the force of vacuum as water

[713] *De Usu Partium*, VII, iii-vi; Daremberg, I, 458 ff.

[714] A piece of the outer edge of this leaf of the manuscript has been torn away; hence eight lines of Harvey's notes on this page are incomplete. The writing on the verso of the leaf, however, is intact.

198

in a reed because: 1st, never is the lung dilated by an im-
mobile . . . with the thorax moving; 2nd, when something
occurs within (as air) not . . . but with the chest pierced
they collapse. **WH** That **X** for I saw a candle blown out,
and . . . and liquid having been introduced perhaps from
another side through . . . or because it binds to the ribs;
thus the wound. . . .

Cadaver with
the chest not
pierced and
liver not
defective **WH**

    WH Likewise 1st **X** Δ in the hysterical: without flat nos-
trils &c.; . . . [According to] Aristotle and Averroes[715]
&c.; that the heart, according to [Du] Laurens [does not
move the lungs];[716] **X** moves the lungs; proportionately
farther in the young and the old; native heat having been
increased, it is necessary to raise the instrument that more
may be accomplished. **WH** The instrument having been
raised the bronchi are extended by inspiration. **WH** On the
other hand, by entrance of cold [air] it becomes smaller,
the vessels are contracted, air is expired. **WH** I believe
rather with Aristotle that the lungs naturally distend the
chest rather than that the chest them,[717] but nevertheless
the chest [does distend] them.

    Since [there are] 2 parts, inspiration and expiration,
whether both [are separate] actions [or] if the actions are
[assisted] by another.

    Because in other animals neither the muscles of the chest
nor of the diaphragm [are moved], and yet birds sing and
modulate, and in others the lungs [are moved] not by the
chest but by the venter, as frogs &c. **WH** Therefore the
lung first and principally has in itself a faculty of moving,
and it assists the chest and diaphragm by a remarkable
agreement [between their movements], as Averroes

---

[715] *P.A.*, 669a:15: "The organ of respiration is the lung. This derives its
motion from the heart; but it is its own large size and spongy texture that
affords amplitude of space for the entrance of the breath. For when the lung
rises up the breath streams in, and is again expelled when the lung collapses.
It has been said that the lung exists as a provision to meet the jumping of the
heart. But this is out of the question." Aristotle compared the lung to a pair
of forge bellows (*see* note to the above passage). Averroes subscribed to
these views.

[716] Du Laurens, p. 376.

[717] *De Respiratione*, ch. 21.

199

[stated].[718] According to Areteus the heart causes the lungs to desire to attract air;[719] the lung has in itself the faculty of respiring.

Of the 4th, [I am in] doubt whether [it is] involuntary or 84v voluntary. The unthinking may believe that the motion of the chest and of the diaphragm [is] merely voluntary because they move about [voluntarily]; thus *flebiteing scratch*; [but consider] in the embryo; it is the way[721] of one sleeping. WH Wherefore in some animals inspiration cannot be accomplished without voluntary action, but [it is] involuntary per se in the case of the intestines, of the uterus, of the heart. Wherefore motion of the lungs [is] necessary for respiration. Motion of the chest [is necessary] for wellbeing and sine qua non. And they occur simultaneously by an astonishing agreement; and while a cough may be a voluntary action, it arises from the [involuntary] expulsive force just as by vomiting and swallowing the latter defend against whatever disturbs the parts. By the natural [i.e., involuntary] faculty of the animal hunger so affects motion, a perceptible object and desire; (Galen [states] that motion depends upon muscles[722]) that no intellectual effort can prevent its arrival. 2nd, the lungs are torpid or collapse; *Chincough*, breath corrupted; they may not follow the motion of the chest; as *by the opening of a pitt choked, as corn heapes in Apuglia, and diing crye O my breath is gone;* by deliberate motion of the chest WH p. 8 *finish;* but that is not done without the animal faculty and thus they are apparently voluntary captives [according to] Galen.[723]

Cicero accuses C. Licinius Macer of extortion &c., ac-

Marginal notes:

Deranged ones Δ Regarding Hill[720]

WH Voice of animal

From Du Laurens p. 373

---

[718] Harvey appears to have taken this remark from Du Laurens, *Historia Anatomica*, p. 376.

[719] *Aretaei . . . de Causis & Signis Acutorum Morborum*, in *Medicae Artis Principes* (Geneva, 1567), bk. 2, ch. 1, col. 7.

[720] The published transcription gives *"Cis:* Hill," which seems more likely to be *"Ci[rca]* Hill."

[721] The published transcription gives *E⁴: Ro*, which seems more likely to be *E[st] R[ati]o*.

[722] This refers only to that division of Galen's motion called transference of position; *De Motu Musculorum*, I, 1; Kühn, IV, 367; Daremberg, II, 321: "The muscles are the organs of voluntary motion."

[723] *De Usu Partium*, VI, ii; VII, 9; Daremberg, I, 380–381, 479–80.

cording to Valerius.[724] But usually arouses such passion as not desirable; *almost can overcvm* Δ Macer's precaution. *Yet these storyes are cited: but they are rare yet sufficient to testefy;* children playing, holding the breath on account of color; until the lungs are torpid [and] unable to respire; *soe curst children by eager crying grow black and suffocated,* the animal faculty not lacking. WH Sometimes the motions of the lungs may be voluntary without consciousness, viz., they swiftly pass into delirium without memory and they may be compressed, as urine.

The concoctive or alterative faculty of the lungs is not 85r understood, wherefore [there is] difference of opinion and doubt.[725] 1, as to whether or not [it has] some common functions; 2, as to the kind of alteration, whether by heating or cooling, and if it is by heating whether [it is] preparatory or concoctive; 3, as to the subject of this alteration and what it does with blood, air and spirit.

This great doubt and uncertainty is a matter which for a long time has required solution. I express an opinion [based on] the arrangement of this body: 1, it appears to have a common function because of the great vessels &c., for nature [does] nothing in vain: common vessels, common use. 2, hence in many excrementa, as the sputum of a cough, which when it is separated appears to have been concocted; on the other hand, in some animals in which the lungs are

[724] Since Harvey states in the marginal note that he took this story from Du Laurens, p. 373, we are able to identify this pagination with the edition published at Frankfurt, 1600. However, the account in Du Laurens was taken from Valerius and makes no mention of Cicero. Consequently, Harvey recalled the statement of Plutarch, although incorrectly, since Cicero was judge, not accuser, in this case. The important part of the incident for Harvey's purpose was the fact that upon being declared guilty, Macer shut his mouth, covered it and his nose with his cloak, and so committed suicide by suffocation.

[725] Harvey spends some time on speculative argument about the functions of the lungs. Respiration, both as to mechanism and physiology, was ill-understood in his time. Richard Lower in his *De Corde* (1669) records how he proved, by insufflating the lungs of a dying dog with bellows, that the blood returning from the lungs became red while the air was being supplied. When deprived of air the blood became dark. It was to be left for Antoine Lavoisier to show in 1775 that this resulted from oxygen being taken up by the blood. He also showed the true nature of the gaseous interchange that takes place in the lungs. Oxygen had been isolated by Joseph Priestley in 1772.

fungous and vesicular, neither blood nor excrementa. Wherefore, if they concoct, as seems true, in certain ones only, viz., where fleshy and full of blood; X in the embryo. Wherefore it does not appear in all that concoction accords with function. Chief function [is that] of motion; and the principal part [concerned is] primarily and principally the bronchi.

Each action according to the school of physicians, 1, it cools and tempers the blood and just as the natural spirit and air are prepared, so vital spirits are manufactured in the heart. Colombo says that he was the discoverer and that the spirits are prepared by continuous motion, for when the blood is agitated it is rendered thin and is mixed with air.[726] It is agitated and is prepared, WH charned[727] by which froth occurs and is apparent of the frothynes of the lungs; (and according to Galen) the spirit concocts the vital air as the fresh of the liver the blood;[728] WH for to what purpose [is] the parenchyma.

Galen [Use of Parts, book] 7 & [chapter] 8.

3, according to physicians vital spirits are made from 85v natural spirits and air; wherefore in this preparation the lungs concoct both. Therefore their excrementa [is] between water and air, a vapor which is breathed out by expiration continually and unceasingly. Therefore inspired, unsuitable air suffocates as [those] ij [suffocated] by burning charcoal: the Emperor Julian,[729] WH the scoller of Cambridg; as [well as] ij who [suffocated] recently in a heated,

[726] Colombo, p. 411: "I add another [use] of importance which has been hitherto unknown. It is the preparation and almost the generation of the vital spirits [in the lungs.] . . . "
[727] The published transcription incorrectly has "crarned." "Charned" is an old spelling for "churned."
[728] De Usu Partium, VII, viii, Kühn, III, 540; Daremberg, I, 475: "It is natural that the external air does not immediately serve and at once nourish the pneuma enclosed in the body of the animal, but that transformed little by little like nourishment, it acquires in time the quality proper to the internal air, and that the first organ of this change is the substance of the lung, as that of the liver . . . is the principle of the transformation of the nourishment of the blood."
[729] Flavius Claudius Julian, "Julian the Apostate," received a mortal wound, A.D. 363, in the course of battle against the Persians. Harvey had apparently heard or read some story, possibly inspired by Christian sources, to the effect that Julian had been suffocated.

plastered room; as in the case of Valerius Quintus Catulus, chosen by Caius Marius to participate in a Cymbrian triumph, [then] chosen for death by Marius, he suffocated himself in a recently plastered room heated [by charcoal].[730]

But more philosophically (not as the spirits have usually been distinguished and separated from the humors and parts as if created and contained in different places), spirit and blood [are] one thing, as serum and whey in milk; and [according to] Aristotle the condition of blood [is] as of warm water; as of smoke and flame,[731] the former the result of the latter; and light from the action of brightness, so that as light [is produced] from a candle so spirit from blood; thus spirit from blood and has its action in becoming as flame of flame.

If spirit were made concocted from air, air as it occurs thinner and thicker; if thick, how [does it pass] from the bronchi into the pulmonary vein; and if the spirit were thinner how would it be generated by the covering of the lungs, when pus and serum [go] through that in empyema.

How (since the mixture is a union of those things altered) could air be mixed and united with blood. That which mixes and alters, if heat, the air becomes thinner; if [the air is] thicker when it is prepared, it is produced by cold, which is impossible in the lungs. Aristotle's powerful argument: if the spirit [is] from air, how [is it produced] in fish which are agile and abound in spirit. Conclusion: opinion WH: those in whom the lungs [are] fleshy and filled with blood concoct blood, since spirit and blood [are] one thing; in the same way as the liver and by the same arguments.

Indeed, rather WH by cooling the breath retains fat and whatever is oily, as one would cool oil *in* [an] *Alembic* or

Spirit not
from air

86r

---

[730] Catulus was the colleague of Marius in the consulship of 102 B.C. In 101 B.C., with Marius, he destroyed the forces of the Cimbri at Vercellae; hence his triumph. However, despite his good relations with Marius at this period, the rapidly shifting political situation in the final period of the Roman republic brought him into opposition to his former colleague, and when proscribed by Marius in 87 B.C., he chose to take his own life.

[731] *Meteorologica*, 384a:13: "those things that are not thickened by cold, but solidified [a form of drying] belong rather to water . . . those thickened [by heat] are made up either of earth or of water and air. . . . Milk and blood, too, are made up of both water and earth."

serpentine, whether balsam or nutritive fat. Wherefore as fleshiness is dried out in the aged, or in disease, of which the sign for Galen [is] beat of the pulse,[732] so the body is dried out and lacks spirit.

Wherefore all affections of the lungs [are] long, induce tabes and [are] incurable; a consumption of all nourishment. Wherefore [according to] Hippocrates height of body in youth [is] adornment [but] a burden in old age,[733] viz., when it is lacking nourishment and spirit in dryness of the lungs. Wherefore according to Aristotle the cause of life and death of youth and of old age is dryness of the lungs.[734]

NB WH if the blood receives coction in the lungs, why does it not cross through the lungs in the embryo.

NB the lungs by subsiding motion propel the blood from the pulmonary artery into the pulmonary vein and thence into the left auricle.

Necessity: there cannot be life without nutriment or nutriment without coction; [nor] coction without heat or heat without ventilation; those who destroy themselves [are destroyed] by exhaustion or by suffocation. Thus native heat [is spent] by cooling and by ventilation, especially by ventilation.

Preëminence: nothing is especially so necessary neither sensation nor aliment. Life and respiration are complementary. [There is] nothing living which does not breathe nor anything breathing which does not live.

The eye having been destroyed, so the vision [is destroyed]; the legs, ambulation; the tongue, speech, &c. If respiration [stops], immediately all [living] things [must die.]

Whether plants which contain warm air swiftly grow to large size

Plants [exist] sufficiently by the entering aliment and surrounding coolness, thus plantanimalia. The mussel [lives] if [it obtains] insects, small flea from surrounding water or air. Fly, WH, respires through its tail; the wasp [is] suffocated in oil. Some things receive the surrounding

[732] *De praesagitione ex pulsu,* in Kühn, Vol. IX, *passim.*
[733] *Aphorisms,* II, 54; Littré, IV, 487.
[734] Cf. n. 708.

matter within themselves. Quadrupeds return the air and whales water by respiring Δ when they breathe out water. Fish give off heat, viz., [from] the blood into the water; and in them the gills are *Jagged*, as the lung in others is divided into multiple bronchi. Hence large animals are hotter and breathe much and often and require greater refrigeration and ventilation; [this is] necessary because they greatly abound in blood and heat. Wherefore all pedestrian things have lungs, in which there is more blood and heat for holding erect the weight of the body. No bronchiatum respires except the newt of Aristotle,[735] wherefore the respiring are the most honorable animals.

Whether there is necessity of air and of its change Whether the odo⟨r⟩ of plants is from their respiration

Why and in wha⟨t⟩ way air is necessar⟨y⟩ for respiring anima⟨ls⟩ and also air is necessary for the candle and for fir⟨e⟩ See WH

Of those respiring, some breathe very often others more 87r rarely. Birds more rarely [because] their lungs [are] spongy and of lesser fleshiness and blood.

Loon, ducks, otter, beaver: rarely breathe but they are much compensated by special qualities; special animals which [breathe] little, breathe often in illness and ventilate themselves frequently; certain ones, such as cetacea, remain under water, yet unless they breathe from time to time they die Δ *porpos*; so fish Δ [in a] *frosty pond*. Frogs and serpents do not breathe in the winter; in the summer *froggs swell*. In some it perhaps suffices to breathe twice in an hour or a day, and that in the summer when more lively and more agile, or when they are ill; so the *porpos agaynst the wether or tempest*.

Δ The sleeping horse very rarely [breathes] by inspiration [but] by expiration.

Worthiness: the lungs [are] the most noble part (with the exception of the heart) as source of blood; without the liver it [i.e., heart] is unable to persist for very long, and without the lung not for a moment. All approaching nourishment is transmitted through the liver, all aliment and the whole mass of the blood incessantly [flows] through the lungs, wherefore arterial blood is redder.[736] The lungs

[735] *H.A.*, 589b:26: "At present only . . the so-called cardylus or water-newt . . . is furnished not with lungs but with gills, but for all that it is a quadruped and fitted for walking on dry land."
[736] Harvey was aware that the blood is red after passing through the lungs. It would seem fair to point out that he did not state that this was owing di-

make the spirits and **WH** indicate the nourishment wherefore more worthy than the liver if honor is judged by
benefit.

Services or use			
Inspiration and Expiration	Agitation of the praecordia	Inspiration	Maintenance of refrigerants, hot matter of the spirit, according to physicians, 'n animal and vital [spirits]
	Motion of the intestines		Propulsion downward of vapors; odors to sensorium
	WH Transit of the blood		drink
		Expiration	Expulsion of sooty vapor Cough and expurgation from thorax Voice, speech Purgation of head by sneezing; driving out of something prickling Clearing, cleansing of nostrils
		Detention	Vomit; expulsion, casting out of urine; birth; allaying of pain; *grone*; augmentation of forces according to Aristotle [IV], p. 665 *frog &c., swine*

Natural and non-natural shapes of heads according to Hippocrates[737] and Galen.[738]

Various authors: Vesalius,[739] Colombo,[740] Fallopius[741]
1. Somewhat depressed on each side without any
    eminence

---

rectly to the respired air, as Lower was later to prove; cf. n. 725. He merely
says that the arterial blood is redder because all aliment goes through the
lungs.

[737] In *Wounds of the Head* (Littré, III, 182) there is mention of a normal
head shape, a slightly elongated sphere, and three variants: 1, prominences
at occiput and synciput; 2, at one or the other; and 3, a perfect sphere.

[738] *De Usu Partium*, IX, xvii; Kühn, III, 751–752; Daremberg, I, 603;
recognizes the same variations of head shape as Hippocrates.

[739] *Fabrica*, p. 16, recognizes the same shapes but adds a fourth variation
in which "the head projects more noticeably at the side of each ear than in
the anterior and posterior regions."

[740] Pp. 35–36: "[The natural shape of the head] resembles a sphere
slightly compressed on each side so that it is slightly extended in length;
you will discover some to vary from this shape, so that they may be now
more pointed at the vertex and more sloping, now with a more prominent
forehead, now measurably contracted, which can equally be said of the occiput. Likewise now more now less compressed at the sides, all which forms,
however, are natural, and they do not differ so much that these may be
called natural and those not natural."

[741] Fallopius, f. 17ʳ, denies Vesalius's fourth variant type: "I agree with
Galen that it is not possible."

2. Increased by an anterior eminence either **WH** in height or width
3. On the contrary, [increased] by a posterior eminence
4. Increased by an eminence on each side in a round head as Thyrsites,[742] according to Homer
5. Increased by a lateral eminence, according to Vesalius rarely,[743] or according to Galen,[744] never; yet it exists

WH And in all there are natural differences, and non-natural according to the greater and lesser width.

The head rests on the first vertebra. According to Galen 88r in a place ordained for the sake of the eyes, but this presumes that use only in man who [is] of erect stature, as those who climb the masts of ships.[745] **WH** Rather it seems that the eyes [are] here for the sake of the brain, and **WH** perhaps for the sake of the eyes and the ears &c.; also for the sake of the brain and its senses; wherefore if we are to be sensible of vision then the eyes must be in the superior and anterior part [of the head], since the most noble of the senses [ought to be] in the most noble place; so in those with no head, as crabs, the eyes [are] in the same place, viz., the most noble [sense] above and in the front; and [there are] extended processes like a head *in a Lobster* wherefore the animal may better search for food and choose a path for flight from harmful things: *snayles* employ touch by horns in place of vision. Wherefore the eyes as *Centinell to the Army* established in an anterior place. [The head has]

[742] As frequently elsewhere, Harvey may well have taken this classical reference indirectly, here from Vesalius (*Fabrica*, p. 16), who, describing his variant form of head, remarks "an exact sphere" and adds, "Homer relates that Thyrsites had this shape of head."

[743] Cf. n. 739.

[744] *De Usu Partium*, IX, xviii; Kühn, III, 754; Daremberg, I, 604: "Another form can be imagined which extends more at the ears than at the front or rear, but does not exist."

[745] On the basis of his doctrine of final causes, Galen explains that the eyes could not be located in a low position, "for their function demands an elevated situation. It is for this reason that those who keep watch for enemies or brigands are posted on walls, on high towers or on mountains. Sailors also, who climb the masts of ships see land sooner than the passengers" (*De Usu Partium*, VIII, v; Kühn, III, 631; Daremberg, I, 538-539).

207

a rounded shape, a little depressed and elongated, WH because nature makes all things round unless for some other
reason; because [it is] the most perfect shape and for the
sake of security. Moreover, nature [is constantly] perfective, wherefore the sky is round and most plants are round;
bones are round. Many other reasons from anatomy for
me X.

WH Among animals those with rounded head more in-
telligent, for [that is] the shape most capacious for the
brain.

Thus the head is large in size; but a small head [indi-
cates] rashness and instability. This arises from its proportion to the thorax, for just as the head or thorax exceeds,
so heat and cold and their consequences. Common meas-
urements: one face from clavicle to ensiform [cartilage].

The head [is] heavy, so that in those suffocated in water 88v
[it is] downward, and those who fall from a height [land]
on the head; so in those in a drunken stupor the head [is]
heavy; [in] children especially,[746] [and in] imbeciles a cold
head [according to Aristotle, vol.] 4, p. 682; [such] are unable to sustain because of cold brain and bones.

Many signs of physiognomy from the head. Large,
small, well-formed, badly formed. Large, small, fleshy,
without flesh. Large, small, well-formed [relative to]
chest, face, neck. According to Avicenna small and well-
formed indicates perfidious, timid, irascible.[747] But large
(unless proportionate to the width of the chest and the bulk
of the neck), mischievous, dull, fatuous. So a badly formed
head, pointed at the top [indicates] a fickle, dull, indocile
and stupid person; well-formed, of moderate size [indi-
cates] a clever, sagacious, astute person.

WH Small [head] on slender neck [indicates] unfortu-
nate, weak and foolish; a short neck [indicates] sensible,
prudent, learned, one who plans; much flesh [indicates]

[746] G.A., 744a:31: "And children for a long time have no control over
their heads on account of the heaviness of the brain."
[747] Canon, Lib. I, fen ii, doct. 3, cap. 5: "If the head and forehead are
rounded and the face is very round and the neck extremely short, and the
motion of the eyes is slow, such men are not of good will."

sluggishness; without flesh, cold and retarded motion; similarly whether or not mischievous, small but better formed. Many other things in books on physiognomy.

The head is divided into the hairless part of the face and the hirsute part of the skull; wherefore to become bald when the hair falls out. Hence in others a large head but small face, so that rounded [indicates] active , *quick nimble-witted*. Others, on the contrary, [may have] a small head but fleshy face with long jaws as a horse.

The synciput [extends] through the forward part of the 89r skull to the coronal suture; this is [indicated if the base of] the hand is placed at the root of the nose and the tips of the fingers in this way [touch] the *Center* of that region. The occiput [extends] to the lambdoidal suture; *Noddle;* the vertex [lies] at the highest point where the hair [is] in a swirl; double [swirl] in some; a sign of future wealth. The temporal [bones are] at the sides, where in time [the hair] begins to become grey.

Parts of the skull. Ordinary containing [parts]: [1] ordinary hair which [has been considered] in its place; [2] skin, epidermis, fat; [3] no panniculus, no flesh. Special containing [parts], external: periosteum, pericranium; internal: pia and dura mater.

Contained [parts]: cerebrum, cerebellum, medulla, veins [and] nerves of the cerebrum and their parts.

Here [there is] hair instead of fat and flesh; preternatural [growth in certain diseases according to] Aristotle,[748] [vol.] 4:680; [in some persons this may] slowly [become]; fat; in others [it surrounds] the cerebrum in a circular fashion; for the brain which [is] cold and very humid and largest in man [requires] the greatest safeguard; for a humid thing [is] easily terminable; by refrigeration, by heating, and therefore it can endure external injuries [only] by reason of so many safeguards.

The pericranium which in some [books is called] periosteum, [is] thin, soft, [with] very acute sensitivity; [originates] from processes of the dura mater [passing] through

---

[748] *H.A.*, 518b:21.

the sutures;[749] excrements are thence carried out. Some hold the same for the periosteum, others distinguish [between the two]; but [according to] Fallopius the same;[750] yet in some at least double as the pia mater; and as [according to Du] Laurens all *Membrane*;[751] or divisible **WH**. [The bones of] the cranium form a double layer with meditullium[752] [between]; [I question] whether as far as the fontanelle; I have seen [the cranium] perforated [by a] *trepan* in which the meditullium was not present. Astonishing how [although] fostering the veins and processes of the cerebrum, it yields to the finger since [it is] soft, as you see; [according to] Hollerius [it is] a flexible body in man.[753]

In some parts [the bone is] thick, in others very thin. 89v **WH** Wherefore [use] caution *in trepanning* [lest] *one side being thorow before the other.* wherefore *not when the brayne swells. coughs.*

WH Without sutures Rule &c. Sutures [act] as breathing devices of the cerebrum; **WH** of expansion; so in the tortoise and for dilatation of the cerebrum; [the question as to whether there are] more [sutures] in the male and more open than in the female. I pass by, but [consideration] of these among the bones.

Since the meninges [are] medium between the very soft cerebrum and the [hard] cranium; a necessary medium and a partaker of two contraries; and since the cerebrum and calvarium are much distant, so the two membranes dura

---

[749] This is not strictly correct. The cranial bones are lined on their inner surfaces with a fibrous membrane, called the endocranium, which is the outer layer of the dura mater. This membrane passes through the various foramina and becomes continuous with the periosteum, or pericranium, on the outer surfaces of the bones. The endocranium is not continuous with the pericranium through the sutures.

[749] Fallopius, f. 216v: "The pericranium should not be differentiated from the periosteum, as some anatomists of our time do, since the pericranium is exactly the same thing in the head as the periosteum elsewhere."

[751] Du Laurens, p. 385: "The pericranium is a dense and solid membrane which since it covers the bone of the cranium externally is called by the special name PERICRANION rather than by the more common PERIOSTION; for I consider that there are not two membranes distinguished here, that is, pericranium and periosteum as many are accustomed to think."

[752] A term used for the diploë. The bones of the cranial vault consist of two compact layers of bone, inner and outer, with the diploë between.

[753] Jacobus Hollerius, *De Morborum Curatione* (Paris, 1565), p. 2.

210

and pia mater [lie between them], the dura always adhering to the cranium; the folded pia unfolding.

And so the dura [mater] is similar to the pleura, the peritoneum, the periosteum; lies in the middle between the pia mater and the cranium. It is as much softer than the cranium as it is thicker than the pia [mater]; it is harder, thicker, denser than all the other [membranes]. Wherefore some believe that it is the origin in the body for all the others and truly the mother [of them].

In its site it enfolds the cerebrum on all sides and is very firmly connected to the cranium, especially in the lower part, and is continuous with the pericranium through the sutures of the head. It is connected to the pia [mater] for the benefit of the vessels. Double like other membranes; as if twin [according to] Fallopius;[754] the exterior part, because of the bone is harder and rougher; is therefore less sensitive. The interior [surface is] smooth, soft, slippery and sensitive. An aqueous humor[755] as a guardian for the cerebrum.

Its parts: veins, vessels, sinus, falx, to assist inflation.

Veins: distended by blood in daughter [of] Dr. Argent; varices in the embryo; wherefore pain of the head mitigated by compression; daughter very young girl, pale, overwhelmed by pain.

Four sinuses like aqueducts enter [the skull] at the base 9or of the cerebrum.[756] [They are formed] here and there by

---

[754] Fallopius, f. 217ʳ: " . . . is very easily divisible into two membranes of which the external is the harder and less woven, and appears like a wide tendon spread out and attached to this membrane. The internal of these membranes seems to be smoother and denser, and the sinuses are present in it as an anatomy clearly shows. Therefore the external portion of this membrane is attached to the bone as a periosteum, but that this membrane is, as it were, double, the majority of anatomists are unaware."

[755] The cerebrospinal fluid.

[756] Harvey's use of the numbers 1 to 4 for the sinuses is best explained by a quotation from Riolan, *A Sure Guide*, trans. N. Culpeper (London, 1671), pp. 121–122. The first and second sinuses are the transverse or sigmoid sinuses of modern terminology, the third sinus is the superior sagittal sinus, and the fourth is the great cerebral vein of Galen continued as the straight sinus. The nates refers to the superior colliculi "corpora quadrigemina," and the testes to the inferior colliculi. "Now the Pipes belonging to this Coat, are four, whereof two are lateral, which run along the sides of the Sutura Lambdoides, that they may receive the blood from the internal

the convolutions of the dura [mater]. The 1st and 2nd [run] upward into the middle; the 3rd and 4th [go] into the testes and nates; wherefore here [according to] Herophilus [lies] his press.[757]

From here many of the veins [extend] by branches in the dura and pia [mater] and cerebrum [and run] here and there; wherefore [there is] an abundance of blood, because [it is] a cold part and uses much nourishment; wherefore it emits much excrement through the nostrils,[758] and the cause of many diseases [arises] from this catarrh.

The 4th sinus [goes] to the testes, where [it undergoes] various divisions, and [passes] into the third ventricle; there it is divided [to go] into the right and the left ventricles; and there they may be joined with the branches of the first and fourth arteries in the formation of the choroid plexus.

The duty of the sinus is in doubt; whether that of vein or artery or of both, for [it is] filled with blood, but it does not have the vein's covering. WH But neither is this so in [the case of] the liver nor of any viscus; but not [as a vein] in function, for [the sinuses] pulsate, whence [occurs] the pulse of the cerebrum. [According to Du] Laurens [there

---

Jugulars, and from the Neck Veins; or by them according to the Doctrine of Circulation, the blood may flow back unto the Heart.

"From the union of the two Channels, is formed a third, longwise, drawn out directly as far as the Nostrils. In the Concourse of the three beneath, there springs a fourth Channel or Pipe, which goes into the Substance of the Brain, between the Brain, and the Petty-Brain; it is not shut up in the foldings of the Dura Mater but there is a great Vein, so called by Galen, which descending into the former Ventricles, makes the Plexus Choroides, which is dispersed through all the Ventricles, unto the Basis of the Brain.

"The Channel which runs longwise, deserves rather the name of Torcular, then the fourth; because from thence is the blood distributed into the lower parts, by innumerable little Veins, through the turnings and windings of the Brain."

[757] Galen, *On Anatomical Procedures*, trans. Charles Singer (Oxford, 1956), p. 228: "The region that Herophilus [is said to] call the torcular. . . . " The torcular Herophili or confluence of the sinuses is the expanded lower end of the superior sagittal sinus which becomes the right transverse sinus after receiving the straight sinus and a branch from the left transverse sinus.

[758] The old teaching that a variety of materials, including blood, could be discharged into the nose from the superior sagittal sinus, is, of course, fallacious.

are] madmen who do not believe this;[759] **WH** I [am] a mad-
man; the pulse arises from artery and spirits, and as a consequence the whole cerebrum appears [to pulsate] because
like a quagmire; as in an indestructible suppurated aposteme;
even if near an artery [it has] the pulse of the artery. For
motion extends into soft bodies; thus it compares to a
cough and convulsion of the voice, and causes the cerebrum to swell. **WH** Hence very delicate pulses *Bob* [in the]
sydes[760] in the ends of the fingers. And so since they pulsate and have blood in place of veins and arteries, this occurs from mixed veins and arteries, and neither artery nor
vein but a sinus for each; and this is peculiar to the cerebrum. Nevertheless Fallopius denies that it occurs from an
artery.[761] The shape inward [is] triangular and in some
place as if two foramina; and it sends branches everywhere. Hence in pain from plenitude the occiput [is] hot.

The dura [is] folded between the cerebrum and cerebel- 90v
lum at the first and second sinuses.[762] Likewise it has a
threefold thickness. The falx dividing the cerebrum into
two parts [is] connected to the septum of the nares[763] [and
is in] the shape of a scythe at the corpus callosum, with
veins for the most part from the third sinus.

It is not [present] in the goose, hare, &c., but the bipartite brain is connected to many slender plexuses. Likewise
it is not found in the sheep's head.

Pia mater, secundina, as corium in the embryo CHO-

By pulse: to
attract air, to
expel sooty
vapor, to make
[animal] spirit

---

[759] Du Laurens, p. 403: "No one, unless he is mad or completely un-
skilled in anatomy, will deny that the cerebrum is moved."

[760] The published transcription gives the seemingly meaningless *babsydes.*
"Bob," a verb meaning "move up and down lightly like a buoyant body on
the water or an elastic body on land; to dance lightly up and down;" it was
used regularly from Chaucer onward of the regular movement of a pendu-
lum, which itself came to be called a "bob." Here Harvey is speaking of
light and delicate pulses, and although the manuscript is very difficult to
read at this point, "*Bob* [in the] *sydes*" is a more likely rendering relative
to the context.

[761] Fallopius, f. 221ʳ·ᵛ: " . . . I would that I could sometime see the motion of contraction and dilatation which Galen was the first to ascribe to
the brain and which some anatomists of our time confirm. . . . I have never
been able to detect perceptible pulsation either in an anatomy on living
animals, in large wounds of the head or from any other sensible evidence."

[762] This refers to the tentorium cerebelli.

[763] This is not anatomically correct, but it used to be thought the falx
had a connection with the nose.

ROIDES. [It is] soft, very thin, just as the dura is very thick. Immediately invests the cerebrum and enters its convolutions. [It has] exquisite sensitivity so that some, [according to] Piccolomini, [consider it] X the organ of the sense of touch.[764] It contains the venules of the cerebrum, as the dura [contains] the larger [veins], and the sinus where it enters the convolutions, which the dura was unable [to do]. It firms and contains the cerebrum as the mesentery fosters the intestines and safeguards like the epiploon.

Similarly both coverings [exist] for the cerebellum and nucha, nerves, entering veins and arteries, so that contrary to Piccolomini the veins would equally carry down sensitivity from the nerves. The cerebrum is the highest body in a very well-protected tower; hair, skin &c., as safeguards, so that nature [protects] no part more; wherefore a principate; nevertheless WH, not to be compared with the heart. Because the empire of the heart extends more widely in those in which [there is] no brain. Perhaps more worthy than the heart, but the heart is necessarily prior, because it is better to be [and to live] well than simply [to live sluggishly]. For this reason since all animals have one most perfect part, man [has] this, excelling all the rest; and through this the rest are dominated; it is dominated by the stars wherefore the head [is] the most divine, and to swear by the head; sacrosanct; to eat [the brain is] execrable.

WH Heat is necessary for sensation as well as for the 91r concoction of food. Since in the beginning of life [there is] greater necessity for concoction of food than for seeking after sensations; for without heat there is torpor; but an animal is distinguished by sensation and [it is] necessary for nourishment. Then it is necessary to assume separate parts according to the different proportion of native heat, [e.g.] heart and brain; wherefore in all sanguineous [animals] in which [there is] very great heat, [there are] a heart and a brain.

[764] *Anatomiae Prelectiones* (Rome, 1586), p. 246.

214

* Wherefore Aristotle, whom X Galen opposes,[765] [declares] the lungs and brain [as a safeguard] because of excess heat, (for heat of heart IV, p. 68) so that it is a structure contrary to the heart and a judge for tempering the heat; wherefore, on the other hand, the nonsanguineous in which the heat is constant, one organ sufficient for both; heart and brain mingled as in earthworms, whether it be heart or brain WH. The hotter the heat in animals the more ample the lungs for blood and the brain for intellectual activity; wherefore man, the hottest of all animals, [has] the largest brain.

* On the Parts [of Animals], IV, 10, the temperament of the blood of that part suitable for tranquility of sensation

Just as in a very small commonwealth [there is] the same judge, king, adviser, so in larger they are separate, and politicians [can acquire] many examples from our art. Just as in the lower venter where [there are] different coctions which require different heats for different preparations of nourishment, so [there are] different organs, in addition to the heart, which [are] furnaces, lares and different operators of different functions. Liver, spleen, stomach [act] as alchemists by means of different furnaces and heats, [for they are] different organs.

Therefore the action of the brain [is one] of sensation, and the brain [is] for the sake of sensation by which animal is distinguished [from plant]; wherefore since [there is] an organ of sensation, different sensible objects and images of the brain are distributed into the nerves; wherefore [according to] Avicenna the nerves [are] like plantings of the brain,[766] [and provide] ready intelligence for the organs of sensation, like WH the fingers of the hand; wherefore the brain neither sees nor hears &c., yet [knows] all things.

Furthermore from reflected sensations you are sensible of that which the sensus communis declares, which [is] always what the nerves and organs of brains [obey]; and it is necessary that it be recognized as the same as seeing, hearing &c. One intelligence for many apprehensibles.

[Intelligence possesses] not only utility, but a kind of re- 91v

[765] *De Usu Partium*, VIII, iii; Daremberg, I, 531.
[766] *Canon*, Lib. I, fen. i, doct. 5, cap. 1.

flection [which] conceives, and from those reflections it forms [ideas], that is phantasy; and [that which] recalls absent things is memory; and one man presents what has been recollected according to his opinion, and since according to his opinion past happenings are diverse, the cogitative sense conjoins them to present events, separates those which he affirms [from those which] he denies; conceives, understands, defines. In affirming and denying is included ratiocination, wherefore a rational mind is especially and properly concerned with this.

Shape [of the brain] as the cranium, *the mowld;* wherein [are] the various convolutions of the head; as vessels enter that X smooth in the goose; in addition veins in the substance of the brain. WH That it is always thus [convoluted] so for the sake of something; but of what I WH do not know; wherefore it does not seem as in Erasistratus that [it is] for the sake of the intellect.[767] For Galen nothing in the part that does not conduce to the action of the intellect; therefore it is a function of the brain; for nature does nothing in vain. Wherefore [according to] Colombo [the convolution provided lightness] for expediting motion.[768] [According to] some it is divided from the cerebellum lest by softness it collapse into the septum. [According to] others so that the vessels might enter. X in the sting ray where they are far apart. Bipartite right and left but here an astonishing phenomenon, for with the right side injured a paralysis on the left, and the contrary. Quantity [of brain] in man largest, the hottest animal;[769] [amount] of two oxen, and more in males than females; weight [according to] Bauhin 4 lbs.;[770] [in] *woodcock* not larger, indeed, than the eye. Not much larger than the medulla. In the tortoise the eye is larger than the middle part;

[767] Galen, *De Usu Partium*, VIII, xiii, Daremberg, I, 563.
[768] Colombo, p. 350, "The reason for these circumvolutions [of the brain] must be sought. I believe it was for lightness. . . . "
[769] *P.A.*, 653a:25: "Of all animals, man has the largest brain in proportion to his size; and it is larger in men than in women. This is because the region of the heart and of the lung is hotter and richer in blood in man than in any other animal; and in men than in women."
[770] Bauhin, p. 572: "It is of such mass that sometimes in weighing it in a public anatomy, it has appeared to equal four and sometimes five pounds."

[so also in] nonsanguineous and cold animals; slight in fish.

Sometimes it is swollen as at full moon; by verbal dispute; [according to] Avicenna damp from moisture at certain times and places.[771]

Its substance [is] very white, very pure as a very limpid spirit; **WH** but all things are black in an obscure place; not more obscure in the timid; very soft in children in whom the reason is imperfect; often fluid in the embryo, soft and sticky like mild and tranquil spirits it is not easily dispersed.

The temperament [of the brain is] cold for [it is] ex- 92r sanguinate, and humid and soft; but cold that it may temper spirits from the heart; lest the heart be inflamed and swiftly lose its power because in the insane the brain has become hot.

Humid lest it be dried out by the motion of the spirits and as is readily apparent from observations; **WH** contrary to the heart, for we said the humor is clear.

The temperament is known from the quantity of the head, length of the neck, the eyes and the quantity and color of the hair &c. Wherefore from these [there are] many signs of physiognomy and health.

Such substance in the rabbit in the cavity of the eye; for cold and slender veins; humid wherefore swiftly *wheyish*, wherefore in penetrating wounds easily putrefies. Wherefore a hot dressing [is used] in treating [wounds]; wherefore it is easily injured in venery for the spirits having first been exhausted by cold, they suffer and endure more from those things which vitiate and are weakened by small and insignificant causes. Wherefore the very humid part easily susceptible to refrigeration and heating; especially suited wherefore: in each way.

All external catarrhs [enter] through the sutures &c.[772]

[771] *Canon*, Lib. III, fen 1, tract. 1, cap. 3: "For in every brain in the beginning of its creation superfluous moisture exists, either from what it acquired from the womb or afterward, which must be purified."

[772] The old teaching that catarrhs, effluvia, and so forth escape from the skull through the sutures was handed down from Aristotle and Galen; it was probably based on the fact that in most animals the cranial sutures close early but in man they do not become obliterated until late in life. Whatever its basis, the teaching was fallacious.

Likewise through nerves, nucha, veins, arteries &c. Wherefore they gush forth as excrement through the palate and nostrils; because [they are] very humid and because also cold to the touch of all [as are] diseases in a superior place. In the form of a cupping glass it receives, collects and attracts vapors, condenses and detains. Wherefore either it liquefies like snow or is compressed like a sponge; distillations may cause many diseases.

Here [is] the seat of sleep for vapor [rises] upward from nourishment through the veins, arteries. Here it condenses and propels downward the heat by which a compression of the senses [with] the head and body powerless for erection.

Temperament is said to concoct spirits, and regarding concoction of spirit 7 opinions enumerated by Bauhin. WH I agree to none for many doubtful, impossible. WH rather blood and spirit for like a flame [with] nothing separable; but these things [will be discussed] more suitably in the anatomy of the nerves.

[There is] doubt as to which is the principal part [of the 92v brain], whether the ventricle or whether the substance. WH The ventricles are seen filled with water: *dasy sheep*; [and act as] a vehicle of water; the waters issue clear; best observed in insects and adequate in the goose; [mention] those in which they are not.

Nevertheless the anatomy of the nerves teaches that sense occurs in the brain, and WH [is shown] by the example of the frog; in cogitations the head suffers, in wakefulness the weak [parts] suffer. Nevertheless in hydrocephalus [according to] Dr. Argent the whole brain is consumed; nothing [left] except meninges; and Hollerius observes that the hairlike vessels of the brain [when blocked by] sanies [are] often in great pain.[773]

WH In Padua in the case of one struck by lightning; use of inunction in lues venerea; alopecia of beard and head. Here the whole skin. One part of the cranium [was] carious, penetrating the meninges by a livid spot; the dural membrane of the brain drawn down, *curded milk*; cancer

[773] Jacobus Hollerius, *De Morborum Curatione* (Paris, 1565), p. 1.

218

size of an egg after flavescent curd; then cerulean callous wherefore hard so that when incised the parts retract from each other; inward, Phoenician purple.

Such as the hard tumor [found according to] Bauhin in Baron Bonacurtio in the year 1582.[774]

The corpus callosum, the upper part of the ventricles, 93r sustains [the weight of] the brain.[775] Two anterior, larger ventricles [which are] semicircular [in shape].

[According to Hieronymus] Mercurialis [it is] in the region of the auricle that the spirits enter with a rush.[776] Within [it is] a shining color WH in the recently dead. Covered with a thin membrane; some contend they are a branch of the *pelvis* [infundibulum]. [According to] physicians when the matter abounding here irritates, epilepsy occurs from vertigo; or, according to others, when the spirits contained here are agitated too much; with the ventricles injured [the brain] is wholly obstructed by apoplexy; wherefore I have seen water in apoplectics and in their plexuses *like chopt brann like a little Poridg;* however, observe in particular the water in those hanged.

NB [in] the brain, with the veins distended with black blood the ventricles are free of water; the cerebellum so putrescent with blood that torn away and touched with the fingers, it will slip about like the distillation of fluid metals, quicksilver.

WH Wherefore I believe that the water here [is] from corruption of the brain after death; especially in youths and in humid bodies, for in the recently dead the brain [is] much more solid.

Here is the manufactory of animal spirits, wherefore by Use moving, the brain [draws] external air through the nos-

---

[774] Crooke, p. 460: "In this place sayth Bauhine, in the year 1582. I found a scirrhus or hard tumor in the noble Baron Bonacurtio (who lay a long time in a manner Apoplecticall or astonished) when we opened his head after his death."

[775] The corpus callosum actually forms the roof of the lateral ventricles.

[776] Varolio, *De Nervis Opticis* (Frankfurt, 1591), p. 157. It was the opinion of Mercurialis that the lateral ventricles had been formed in irregular rather than circular shape so that the irregularities might slow down the forceful movement of the spirits into them.

trils[777] &c.; wherefore there is a plexus of veins and arteries; Fallopius [believes the plexus is] from the arteries alone.[778] [According to] others the motion [is] from the ventricles, [and] intelligence [resides] in the substance of the brain; wherefore [according to] Hippocrates if disease has crept into the mind, [it is] melancholy; if otherwise, viz., into the ventricles, epilepsy.[779] Some [believe it is] a promptuary of the spirits, wherefore [according to] Galen two for the sake of security; the one injured in the case of the youth of Smyrna.[780] Some [believe] the cloaca of Piccolomini for excrement and purgation; wherefore others who attribute to the ventricles [the functions of] fantasy, first [power] of the intellect &c. They place the soul in the workhouse, they mix the most divine faculty with excrements superior and inferior; others that the choroid plexus is wholly glandular to touch. [According to] Bauhin, whence Varolio, its use is to absorb the humors of the glands;[781] WH I believe it is a proc-

[777] Riolan states that the air drawn in through the nostrils does not enter the lateral ventricles "because they are void of any inlets." Therefore the air does not mix with the spirits in the ventricles but passes over the outside of the brain to cool its surface.

[778] Fallopius, f. 133ʳˑᵛ: "A canal is formed for this artery [middle meningeal a.] in the dural membrane through which it is distributed by means of numerous shoots, and this is the reason why this membrane is seen to pulsate. The final residuum of this third branch creeping through the temple near the anterior root of the ear and the maxillary joint, reaches to the side of the forehead and there terminates. After the third and very significant branch has been distributed, the whole of the remaining artery, which is little less than half the trunk of the whole soporalis [common carotid a.], although it is sometimes the larger part of it, ascends but remains intact. It reaches the special foramen carved in the petrous bone and ascends through it into the cavity of the skull to the aforementioned sella where, while hidden under the dural membrane, it gives off an infinite number of small shoots and forms the rete mirabile in brute animals."

[779] Hippocrates, *Epidemics*, bk. 6, sec. 8, par. 31; Littré, V, 357.

[780] Galen, *De Usu Partium*, VIII, x; Kühn, III, 663–664; Daremberg, I, 557: " . . . at Smyrna . . . I saw a youth injured in one of the anterior ventricles survive this accident. . . . It is certain that he could not have survived a moment if both had been injured simultaneously."

[781] Bauhin, p. 699: "The use of this is seen to be like that of other glands. They consume the superfluous moisture of the brain. Hippocrates judges that there are glands in all parts of the body to free it from superfluous moisture . . . indeed, he decided that the whole substance of the brain is such." Varolio, *Anatomia* (Frankfurt, 1591), p. 9: "Two notable cavities were formed in the cerebrum which are called ventricles, into which such kinds of excrement are deposited as into a sewer. It is said that it is for this

ess of the pia [mater] for the plexus is everywhere in the cerebrum.

WH But in the goose, in the duck &c.: they are not en- tirely [small] although more in the rabbit, mare; and in the goose [there are] cavities in the testes; so *in a playse* and I believe in almost all fish and fowl; WH and where there are quasi foldings of the brain, they are rather of the medulla. These two join and in truth form only one, wherefore in some only 3 ventricles, in others 6, in others 4 and in others 2.

Testes and nates;[782] the superior testes, inferior nates under the fornix, in truth outside the cerebrum in the inferior and posterior part of the passage from the third to the fourth [ventricle] where the cerebellum having been removed a foramen called the anus;[783] a long pit between them, [which according to] Colombo [is called] the vulva;[784] in man [it serves] to carry down the vessels of the fourth sinus into the cerebrum and to sustain the cerebrum from compression of the third ventricle.

Anus
vulva
use

The pineal gland where it enters the 4th sinus [is] joined to the choroid plexus; is called the penis from a likeness corresponding to the testes. It closes the orifice of the ventricle and sustains the vessels, wherefore [it is] under the vessels.

Testes much larger; joined in the cerebrum and cerebellum so that they are more distant in the sting ray and almost all fish. WH Thus *in a Ratt* a very long interval between the cerebrum and cerebellum; WH testes on the contrary, or at least the process, in goose and hen [are] hollowed: nates obscurely in the same receptacle.

Third ventricle [from] the concurrence of the first two;

---

reason that certain numerous unclean and loose-textured glands, formed of membranes like sponges, suck out from those sinuses the excrement transmitted there through vessels from the cerebrum and cerebellum."

[782] The nates and testes were names applied to the superior and inferior colliculi respectively. Harvey is confused when he calls the testes superior.

[783] Refers to the aqueduct into the fourth ventricle.

[784] Colombo, p. 354: "Furthermore in the anterior part of the testes, extending toward the third ventricle, you will find another part of the brain which not inelegantly and very precisely presents the appearance of the female pubis and vulva."

some deny that there is a ventricle here, so that according to them [there are] only two ventricles. I call the long fissure of the cerebrum the center of the cerebrum; [it has] a double cavity, one ending in the infundibulum, the other under and behind the testes and afterward extending more anteriorly to the passage of the former into the infundibulum: wherefore: in place of cloaca of the ventricle the fornix ▷ a longish triangle anteriorly pointed, a vaulted body in the anterior angle; in the extremity of the septum lucidum a line from the middle of the protuberance continuous with the septum. Immediate through the plexuses.

A candle placed behind the septum lucidum [causes it to resemble] the host of the mass[785] or lanthorn; [it is] a membranous portion of the brain.

The 4th ventricle [is continuous with] the origin of the spinal medulla; the calamus scriptorii [lies] between the cerebellum and spinal medulla; it is wrapped in a thin meninx. Some deny that it is a ventricle. [There are] those who dispose the faculties according to the ventricles, like [Michael] Scot, Thomas [Aquinas], Albertus [Magnus] &c., locate memory here. [There are] those who are seen to refute the idea of the ventricle as a cloaca, because [there is] no passage into the infundibulum by which an exit may be made.

The cerebellum with the brain [has] the appearance of a French fleur de lis; *red*.

[The cerebellum has] right, left and middle parts; two globes behind one another; in the sting ray divided into two; from the higher middle part of the vermiform processes before the nates rearward in the fourth ventricle; when they are retracted, open, relax and obstruct.

WH Wherefore their duty does not appear to be as doorkeepers because nothing acts by relaxation; likewise nucha in the processes in which they move apart; pons

---

[785] It is interesting to note that Riolan refers to this structure as the speculum lucidum. In the translation by Culpeper it is called "speculum lucidum, or the Bright Mirror, because it is transparent." Du Laurens (1600) refers to it as "septum lucidum, speculum lucidum, lapis specularis." Vesalius, *Fabrica*, p. 544: "A more suitable comparison might be a moist wafer such as mass-priests employ."

cerebelli of Varolio. Vermiform processes. Membranous, wherefore [according to] Piccolomini they are not parts of the cerebrum.

Its size [is] four times or ten times less than the brain.[786] Use: it seems almost a brain because almost the same substance; it is of a more pale color; [according to] Plater [it is arranged] in the form of coils,[787] but **WH**, the brain somewhat harder and more compact structure, wherefore reasonable [to think it different from the cerebrum]. If functions of the memory in the substance of the brain, as *conserving* better.

[According to] Galen for the motor nerves;[788] wherefore [according to] some the cerebrum for the senses, the nucha for motion, wherefore by obstruction of the 4th [ventricle] paralysis occurs; the anterior obstructed [leads to] coma of motion and sensation; wherefore somnambulist concerns only the anterior obstructed.

[According to] some, motion and sensation from the cerebral nerves, but sensation from the soft, motion from the hard. **WH** But this is false respecting the recurrent nerves.

Then [let us discuss] the arrangements of the anasto- 94v moses of the nerves. For the nerves only carry down, they do not act, move or are they sentient by a faculty, but [are] organs. [According to] Varolio the cerebellum [is] for the auditory sense and the cerebrum for the visual,[789] but the optic nerves terminate in the brain, the acoustic in the pons of the cerebellum; and therefore dryer and more solid. Because [it is] always thus, [it is] the opinion of Fernel, as before [it was that] of Piccolomini,[790] that motion [is derived] from the cerebrum and sensation from the meninges; wherefore the nerves feel pain of the head, and the meditullium[791] [is] the motive faculty to the tunics which [are]

---

[786] *Fabrica*, p. 540: "The cerebellum is a tenth or eleventh part of the cerebrum." Fallopius, *Institutiones Anatomicae*, in *Opera Omnia* (1606), I, 30: "[The cerebellum] is four times smaller than the cerebrum."

[787] Plater, p. 181.

[788] *De Usu Partium*, VIII, vi.

[789] *Anatomia* (Frankfurt, 1591), pp. 14, 25.

[790] *Anatomiae Praelectiones* (Rome, 1586), p. 262.

[791] The diploë of the vault of the skull.

from the sensitive meninx. WH But I can display nerves from the brain to the meninx. Nor is it sufficient that the brain move in that way so that it acts as a neutral origin. Nerves from the brain [carry] sensation and motion to the muscles and organs; some of which perhaps are motor from the medulla, because the nucha[792] [is] hotter [according to] Aristotle because of fat; and a special one to the heart.

WH And to move is active as to feel is passive, wherefore by greater heat; whether the faculty [of sensation combines] with the essence, viz., of the pertransient spirit, [according to] Galen when [he says the faculty is in the brain] without the essence.[793] I believe that the spirit does not advance by way of the nerves but occurs like a ray of light or an impulse, wherefore sensation and motion [are] as light in the air; perhaps [as] in the flux and reflux of the sea. Native heat immediately underlies the spirits, wherefore [with the spirits] withdrawn or because of intemperateness or impacted humors the medulla of the nerves [is] adumbrated; resolution occurs when the essence impedes the faculty; an accumulation without essence. For the spirit because [it is] the same, [is] unlike a collection; WH With heat so that [according to] Aristotle the instrument of instruments as the hammer;[794] and so the nutritive spirits [arise] from the ventricles of the brain. All things are moved by a spirit, both brain and heat; wherefore the spirited [are affected] by playfulness; do not know how to remain still.

Rightly, Galen, you boast of a discovery unknown to 95r Aristotle, for when Aristotle asserted that the nerves arise in the heart, he referred to the fibers which correspond pro-

[792] This usually means the back of the neck, but the term was used by Alessandro Achillini (1463-1512) in his *Anatomicae Annotationes* (Bologna, 1520) to refer to the spinal cord and the spinal nerves. It would seem that Harvey is also using the term in this sense.

[793] *De Placitis Hippocratis et Platonis*, VII, 3; Kühn, V, 600 ff.

[794] *G.A.*, 789b:8: "So it is reasonable that Nature should perform most of her operations using the breath as an instrument, for as some instruments serve many uses in the arts, e.g., the hammer and anvil in the smith's art, so does the breath in the living things formed by Nature."

portionately to the tendons, the organs of motion which Aristotle also called nerves.[795]

Diversity [of opinion] about the number [of nerves] provided; some [say], 7, 8, 9; mamillary processes and olfactory passages.

1 Some [declare that the fibers] of the optic nerves [are arranged] in [the form of] a cross, wherefore the object does not appear [as] 2 and the eyes are moved simultaneously. For the sake of strength [because of their length]. Some [declare] that the spirits make a transit [from one to the other because] with one eye closed [there is] motion only in the one [open]. Many opinions regarding the juncture which Vesalius refutes by three examples; for in some they are not [as in] *ratts*, nor in some men. Obstructed here [in] the blind; *Catarrack*.[797] WH I have known and have cured from ulcers of the nares; here Plater [describes a

Riolan the optic cavities in the recently dead[796]

---

[796] *P.A.*, 647a:25: "Again, as the sensory faculty, the motor faculty and the nutritive faculty are all lodged in one and the same part of the body. . . . it is necessary that the part which is the primary seat of these principles shall on the one hand, in its character of general sensory recipient, be one of the simple parts; and on the other hand shall, in its motor and active character, be one of the heterogeneous parts. For this reason it is the heart which in sanguineous animals constitutes this central part."

[796] Riolan, pp. 393–394: "Vesalius saw the optic nerves naturally separated without any injury to vision, and Cesalpino notes of a public dissection performed at Pisa in 1590 that one of the optic nerves was found attenuated, the other of full size; but there was a weakness in the eye to which the attenuated nerve was carried. The cadaver had had a wound in that region of the head. However, the attenuated nerve did not extend to the opposite side but was bent to the same one, wherefore all the spectators considered that this proved that the optic nerves never interesect one another but are joined and return to the same side. Galen, *De Usu Partium*, bk. 10, wrote that the optic nerves are hollow in order to carry animal spirits, although in recent carefully inspected dissections no hollow has been found. Vesalius employed the greatest effort in investigation of the hollow of the optic nerve and finally decided that the nerve is solid, filled with various fibers mixed with a medullary substance. Those who defend the hollowness of the optic nerve say that it is plainly apparent in apes and also in man, but the nerve must not be cut transversely nor immediately at its origin, but a little before its insertion. However, Galen requires no excuse or defense, since he explained his opinion in *De Placitis Hippocratis et Platonis*, bk. 7, ch. 4, and in his little book *De Oculis*, ch. 3, for he requires three considerations for observing the canal of the optic nerve: that the animal be large, that it have been recently slaughtered, and be observed in the open air. Furthermore, if the nerves are not hollow, how does amaurosis occur. I have seen optic nerves to be hollow when cut transversely, and the canal appears before the junction of the nerves."

[797] A cataract.

225

case in which the optic nerves were] compressed by a scir-rhus tumor;[798] *the black Catarack*, the gutta serena[799] of the Arabs *vncouchable*; wherefore the use of [the nerves] appears.

infundibulum

WH These [carotid arteries] dried out and retracted, as in wakefulness in old age, wherefore cavities of eyes retracted inwardly; a sign of dryness of brain in tabes and coitus; on the other hand [those that are] humid and relaxed [have] prominent eyes, so that by defluxions physiognomists consider certain ones as tenderhearted, compassionate, for humid [eyes] easily weeping; near entrance of carotid artery, so that the [animal] spirit [is readily available].

Carotid [arteries] to the ventricle follow fifth vein from the dura [according to] Vesalius[800]

Second pair [of nerves] *what on the one side* by an image on the other. From the sides of the base of the brain lesser ones to the special foramen in the sphenoid bone.[801] Long foramen [for] the exit of nerves, viz., these and [also] the third pair.[802] Either the said branch of the third pair and the eighth pair [have] another [foramen], or [it is] for the lesser root of the fifth pair moving the eyes. One branch of the third [pair] which per se rather that it takes origin and exit from the conjugation in the lowest and posterior seat of the medulla. From the lowest and posterior seat (WH inward from the 3rd) by a slender nervule[803] and straight under the base of the brain; forward at the sides and alone perforates the thick meninx and extended with the 2nd pair.[804] Common foramen.

A second branch of the third [runs] to the side,[805] [but] 95v

[798] *Praxeos Medicinae* (Basel, 1625), I, coll. 170 ff.: "De visus laesione."
[799] A term applied to amaurosis, a partial or total blindness with an enlarged and fixed pupil.
[800] *Fabrica*, p. 327.
[801] Probably refers to the oculomotor nerve (second pair of early terminology) and the sixth (abducent) nerve.
[802] The third and fourth pairs of cranial nerves were the trigeminal nerve of modern terminology. Here Harvey refers to the ophthalmic division of the trigeminal.
[803] Refers to the trochlear nerve.
[804] The oculomotor nerves.
[805] Possibly refers to the motor division of the trigeminal nerve leaving the skull through the foramen ovale.

so near the fourth that it seems a branch of it; and it is mixed with the fourth in a common foramen; the third and fourth for taste.

The fourth differs somewhat in origin from the third, [but] not in its course; it arises to the rear of the second pair at the origin of the spinal medulla; [according to] Piccolomini under the fourth ventricle, [806] [according to Du] Laurens below the cerebellum. [807] From its origin [it is] twisted like a goat's horn and is united with the 5th [or] auditory by two ramules; [808] it makes exit through the foramen of the 6th, of the sphenoid bone; it arises by a double shoot which in the same place perforates the thick meninx, wherefore properly [it is] one pair since it is contained by one exterior tunic; it does not permit the second pair and the third pair in the same foramen.

The 5th pair, acoustic, auditory, [arises] from a double origin, the one soft and the other hard; they are covered by the same tunic of the dura mater to the petrous bone. [809] These in fish; wherefore WH the auditory organ [exists] in the sting ray.

6th pair [810] by the same double nerve perforating the dura mater [runs] to the second foramen of the occiput where the jugular veins enter and make sinuses.

The 7th slips from the medulla in the cranium by many roots distant from each other, [811] [extends] obliquely and by a special foramen in the occiput. Here Acquapendente [described] a nerve from the brain into the dura; whether one to the dura.

Ratt: third pair arises before the first; in the goose it is joined to the optic [nerve] in exit. In the goose long nerves under the calvarium from the posterior beak, for what use I do not know. Such in the mole. *Ratts.*

---

[806] *Anatomicae Praelectiones* (Rome, 1586), p. 263.

[807] Du Laurens, p. 109.

[808] The fifth pair comprised the facial and auditory nerves of modern terminology.

[809] The facial and auditory nerves enter the internal auditory meatus.

[810] The glossopharyngeal, vagus, and accessory nerves of modern terminology.

[811] The hypoglossal nerve.

WH NB of the nerves	Foramina	Long [one] near exit of optic [nerve] for 2nd [oculomotor] and shoot of 3rd [ophthalmic division of trigeminal]
		For 4th, third and 4th through three [foramina] in base of skull
Origin	Of second, later of 4th, of third, 4th	[One near jugular] sinus for 6th [glossopharyngeal, vagus and accessory]
		Special [one] for 7th [hypoglossal]

The pelvis[812] [is so called] from its similarity [to a cup]; 96r it arises from a wide orifice, afterward more compressed at the pituitary gland, from the pia mater at the third ventricle.

The pituitary gland [is] outside the thick meninx of the sphenoid bone placed at the end of the infundibulum; above [it is] concave, below gibbous, a small quadrate more compact and harder than other glands; all of it is covered by the thin meninx, wherefore Colombo believes that the thin meninx arises here, and is astonished that Vesalius did not know it; I am astonished that he should believe[813] its use [is for] straining the pituitous excrement into the palate; the bone is perforated by a double foramen at the side by which the second pair of nerves issue; it is led by a second foramen more to the posterior and through a rough fissure. At the sides of the foramen a noteworthy branch of the carotid is transmitted into the skull.

The rete mirabile is declared by Galen[814] [to be] the factory of animal spirit; a mixture of the carotid artery with veins;[815] they surround the pituitary gland everywhere. From the sides of the sphenoid bone like a net; [from] above [like] the net of a fisherman, whose foldings connected to foldings. Bauhin, contrary to [the opinion of]

---

[812] The infundibulum of the pituitary gland.

[813] The published transcription erroneously gives *cognoscerent* instead of *cognosceret*.

[814] The published transcription erroneously gives *de tanta Galenus* instead of *declaratur a Galeno*.

[815] This refers to the cavernous sinus on either side of the body of the sphenoid and the pituitary fossa. The rete mirabile does not exist in man. Galen's description of it is based on the dissection of animals.

Vesalius, [writes that] it is a capital organ in men,[816] but it is apparent in cattle &c. [According to] Riolan with the dural meninx removed [is] widespread and another rete in the base of the cerebrum from fibers of veins, as the former rete of arteries.[817]

In the second dissection

All the nerves [arise] from the medulla [according to] Bauhin's illustration. The spinal medulla arises from four processes,[818] two larger from the brain in front, the others smaller [and] shorter from the cerebellum. The front ones bend away from each other in the middle just where the optic [nerves] come out from the ventricle.

The nucha [arises] from these [according to] Avicenna, vicar of the brain

four eminences of the brain[819]	2 corresponding to the frontal bone
	2 to the cavity of the sphenoid bone

Origin of the optic nerves lying hidden between the brain and the origin of the spinal [nerves]; from origin of the spinal medulla, wherefore cautery to the occiput is of value for the eyes. NB *remove* the pia mater from the intact nerves.

Connection; origin of the brain almost contiguous to the spinal medulla. Course of the olfactory passage to the posterior [is] more attenuated as far as the lateral extremities of the brain in the region which [is] above the auditory foramina. [Advances] into a very acute point, wherefore consensus of auditory and olfactory [nerves]. In head cold, with breathing affected, the air [becomes] apparent [i.e., hearing is affected]. [According to] Fallopius [is observed] to be perforated in a fresh head[820] (which X Va-

[816] Bauhin, p. 609.
[817] Riolan, p. 389.
[818] The cerebral and cerebellar peduncles.
[819] The frontal and temporal lobes of the brain.
[820] Fallopius, ff. 137ʳ·ᵛ: "In animals, such as cattle, goats, sheep and other similar forms, it is by no means difficult to see with open eyes that these processes arise from the anterior extremities of the ventricles and also that a foramen obviously extends from the ventricle into each process. The processes have a passage, large or small in proportion to the process, leading from this foramen to the os colatorium [Ethmoid]. In the bovine the process

229

rolio opposes) [821] from the ventricles; WH Δ in the head of a calf; he believes that first entering as [pure] air it issues as musty; its site [lies] between the thick and thin meninges, first part of the ventricle of the hippocampus of Aranzi [822] and the very wide plexus of the membrane; plexus [is] everywhere part of the pia [mater]; two foramina in the infundibulum; third ventricle: fornix; fourth ventricle.

Anus. [823] Space [made] by contact of the four trunks of spinal medulla, formed of medulla of the trunk of the brain and of the cerebellum; testes and nates [are] portions of the medulla [extending out of the cerebellum]; in truth outside the brain; WH [portions] of the nerves from the underlying parts to the pia mater.

The process of the cerebellum enfolds the spinal medulla 97r as the muscles of the larynx. Varolio's pons cerebelli; from this process [arises] the auditory nerve; wherefore [according to] Varolio the cerebellum is used for hearing. [824]

---

is large, but in the human it is very slender; indeed, so slender that unless examined in a very fresh and undamaged brain, it can never be disclosed. This perhaps was the reason that account of these processes was quite unknown to other anatomists. I believe . . . that air and vapors are carried through these processes and their channels to the ventricles of the brain."

[821] Varolio, *Anatomiae*, (Frankfurt, 1591), pp. 23–25.

[822] *Anatomicae Observationes* (Venice, 1587), p. 43 (Cap. I: "On the ventricles of the brain named from the hippocampus"): "In addition to the already observed sinuses in the substance of the brain, which we have usually called ventricles, in which the very pure animal spirits are made from air inspired through the olfactory organs and vital spirit carried through the carotid arteries, I find two other notables sinuses or cavities hidden and deeply concealed in the deeper parts of the brain, which are not much smaller in size than the upper sinuses or ventricles; and they are circumscribed by a membranous and very solid substance of the brain like the former. They exist under those two anterior ventricles, and they are hidden here and there, as if in an underlying hidden chamber of some ship, and they extend anteriorly toward the forehead and issue connected to a third, or so to speak, common sinus, just as the two superior ones issue joined together and meet with that one as at the center of the brain."

[823] Refers to the aqueduct of the fourth ventricle. The roof of the fourth ventricle is formed of the superior cerebellar peduncles and the anterior medullary velum with, lower down, the tela chorioidea.

[824] Varolio, *Anatomiae* (Frankfurt, 1591), p. 26: "Just as the cerebrum produces from itself the notable trunk of the spinal medulla, from which nerves serving the eyes then arise, so the cerebellum first gives off a large transverse shoot which I call the bridge of the cerebellum. From this emerge the auditory nerves. If one were to say that the conjunction of the auditory nerves in the bridge of the cerebellum is the reason that sound may be heard by one and not necessarily by another. . . . I would not disagree."

Hence [according to] Varolio not all nerves from the brain or from the spinal medulla; for this from the cerebellum; vermiform processes.

No ventricle in the cerebellum; only a sinus in that part which looks back to the nucha; on the other hand, **WH** sometimes in some animals. Aranzi calls it the cistern.[825]

In the second dissection

The brain of a chicken as *a Toad*. Arma Gallinae in the goose as in winged things; **WH** I discovered hollow testes; inward, with a callous body inward, smooth and white.

Appendix **WH** of the Nerves
encountered in the eyes during dissection

---

1 The optic nerves, first [do] not [pass] through the dura mater, [but go] by their foramen.

2 The motor [nerves] called the second pair; contrarily through the dura mater through the long foramen[826] into the orbit.

3 The lateral [nerves], *stradling synews;* according to some the root of the fifth pair, according to others the lesser root of the third and fourth.[827] [According to] Fallopius the eighth pair [pass] through the dura mater. [According to] Fallopius lie hidden in the dura mater; through the long foramen [according to] Fallopius; the whole into the trochlear muscle [according to] Fallopius.[828]

[825] The posterior portion of the fourth ventricle? Aranzi, *Anatomicae Observationes* (Venice, 1587), ch. 7: "On the cavity of the cerebellum which we call the cistern."
[826] The superior orbital fissure.
[827] The ophthalmic division of the trigeminal.
[828] The trochlear muscle is the superior oblique of modern terminology. It is supplied by the trochlear or fourth cranial nerve, which was not clearly defined as a separate nerve in Harvey's time although it had been described by Achillini in 1520. Fallopius, f. 155ᵛ: "Therefore the eighth pair [trochlear] arises as a slender nerve from the lowest and posterior aspect or base of those processes of the brain which are called the nates by anatomists. After arising here the nervule reaches the side of the base of the brain, and hidden in the dural membrane is carried anteriorly along the region through which the second [oculomotor], third [trigeminal] and fourth [abducens] pairs of nerves emerge from the dural membrane. And so this nervule, lying hidden in that membrane, reaches the foramen of the second pair and passing

4 Of the mouth and palate [829] *Mowth: synews*. Taste &c., very thick. [According to] Fallopius the third pair, [830] [according to] Bauhin the fourth pair for tasting; [831] through the dura mater in its reduplication above the spina Lythoides; [832] through two foramina in the sphenoid; above into the palate at the root of the winged process; below into the muscle lying hidden [in the mouth; according to] Fallopius into the tongue and lower jaw. [833]

5 Internal ones; [834] *close long sinews*. [According to] Fallopius the lesser branch of the fourth, [835] according to Vesalius of the fifth pair, [836] [according to] Bauhin of the third pair. [837] Arisen from the middle of the base of the cerebrum

into the cavity containing the eye, is entirely inserted into the fifth muscle placed in the internal angle which, reflected at the cartilaginous pulley, draws the eye circularly to the internal angle."

[829] In this paragraph Harvey refers to the maxillary and mandibular divisions of the trigeminal nerve.

[830] Fallopius, f. 139ᵛ: "I differ by the width of the sky from the rest of anatomists on the third pair, since the nerve which they regard as two pairs, that is to say, the third and fourth, I count as only a single pair. Although the origin of this nerve seems to be formed from many chords, some softer and others much harder, as it is single, arises from a single region and perforates the dural membrane only at a single site, it is not such that we may divide it."

[831] Bauhin, p. 654: "The fourth pair of the tongue, sensory NEURŌN GEUSTIKŌN or gustatory because the sense of taste occurs through them."

[832] The ganglion of the trigeminal nerve passes into a dural sac beneath the tentorium cerebelli just lateral to the tip of the petrous temporal bone.

[833] Fallopius, ff. 145ᵛ–45ʳ: "That called the fourth [inferior alveolar n.] is larger than the other three and descends through the external surface of the hidden muscle [external pterygoid]. It reaches the internal foramen of the jaw and, as it is disseminated through this opening, it approaches the roots of the lower teeth. This portion of the nerve slips through the external foramen of the jaw to the muscles, lips and skin lying there. It should be noted, however, that before the fourth nerve enters the jaw it usually gives rise to a sizeable nervule [mylohyoid] which is very long and slender and which creeps beneath the jaw until it is inserted into the fourth muscle opening the jaw [diagastric] as well as into those placed beneath the chin, and inserted partly into the hyoid and partly into the tongue."

[834] The fifth pair are the facial and auditory nerves of modern terminology.

[835] Fallopius, f. 147ᵛ: "I have placed as fourth [abducens] that which other anatomists call the lesser shoot of the fifth, so that a vacancy will not occur between the third and fifth pairs."

[836] *Fabrica*, p. 368.

[837] Bauhin, p. 658: "The very narrow and twisted inferior foramen gives forth that which carried transversely above the masseter muscles is then mixed with a branch of the third pair resembling a twisted goat's horn, or shoot of the nerve, which is carried to the tongue, and from the tongue is

and a little under, creeping in the brain through the dura mater; a little below and interiorly; from the nerves of the mouth, through the long foramen [according to] Fallopius; [838] WH their own [foramen]. [According to] Fallopius the whole into the indignatorius muscle. [839] WH They are seen joined with those of tasting beyond the dura [mater]. [840]

98r

	Above the optic in	raising the eyelid above the eyelashes
By branches of the eye NB These divisions from Bauhin By the 3rd pair or lateral	Under the optic in	a toper with many offspring Insignificant by bifid branch and several little fibers and by other fibers. Inferior amatory [muscle][841]
	Into the membrane of the eye and whether or not into the innermost NB Fallopius Observationes 224 [842]	

| p. 227 [844] 2nd of Fallopius by branch of 3rd through | Long [shoot] which [goes] with the rest of the nerves and the attached vessels | Larger [shoot] above 2nd pair [843] through superior part of the eye between periosteum and fat Smaller &c | By the exterior angle to forehead In the internal angle in M[an] closing the eye and [to] the muscle of the forehead |
| | Superior Inferior into tongue | ——————— | ———See Fallopius 228 [845] &c |

carried to the buccal muscles and to the skin around the root of the auricle; perhaps it confers no less taste than hearing; which perhaps is the reason that those born deaf are usually also mute."

[838] Fallopius, ff. 147ᵛ-148ʳ: "This pair arises deeply from the middle of the base of the brain; as it were, from the origin of the spinal medulla. Creeping for some distance beneath the brain in an anterior and lateral direction, it perforates the membrane between the second and third pairs mentioned and then sometimes joined with or sometimes without an arteriole, it enters through the foramen of the second pair into the orbit of the eye."

[839] Ibid., f. 148ʳ: "It is inserted almost entirely into the muscle which draws the eye straight to the external angle and possesses no evident relationship to the temporal or to the hidden [pterygoid] muscle." The indignatorius or scornful muscle is the lateral rectus of the eyeball. It is supplied by the abducent or sixth cranial nerve. The abducent nerve was originally described by Eustachius.

233

[440] The chorda tympani is a branch of the facial which joins the lingual nerve outside the skull. It is secretomotor to the submandibular and sublingual salivary glands, and also supplies taste fibers to the anterior two-thirds of the tongue. The chorda tympani nerve was described by Fallopius.

[441] The superior and inferior oblique muscles of the eye were called the amatory or ogling muscles because they rotated the eyeball. The inferior oblique is supplied by the oculomotor nerve.

[442] *Observationes Anatomicae* (Cologne, 1562), p. 224: "This is the correct distribution of the second pair which does not extend to all the ocular muscles but only to four of them and to that which elevates the eyelid. It does not transmit shoots to the temporal muscles as some have imagined. But observe especially that since there is so much confusion in the descriptions of others, unless very great care and much time are devoted to the dissection, the nerve will appear to descend not only to the temporal muscle but also to the teeth. Nevertheless, if one dissects with patience, capability and skill, he will find these features which I have described to be entirely true. . . . In regard to the third pair, I differ, from other anatomists by the width of the sky since that nerve which they regard as two pairs, that is to say, the third and fourth, I consider as only a single pair. Although the origin of this nerve seems to be formed from many cords, some softer and others much harder, since it is single, arises from a single region and perforates the dural membrane only at a single site, it may not be divided."

[443] Probably refers to the frontal nerve, a branch of the ophthalmic division of the trigeminal.

[444] Fallopius, *Observationes Anatomicae* (Cologne, 1562), p. 227: "the third pair [trigeminal] arises beyond the second [oculomotor] from a rather extensive origin at almost the middle of the base of the brain. Consisting of a conjunction of many parts, it inclines toward the side of the bone containing the gland [hypophysis]. By a single perforation of the dural membrane it advances almost into the middle of the retiform plexus in those animals in which this structure is present, or into the conjunction of those two sinuses of the dural membrane which are united at the sella in the base of the head. Outside the membrane this pair immediately divides into three shoots. The first of these [ophthalmic division of V] is that which is directed anteriorly with the second pair, the fourth [abducens], and the eighth [trochlear] or, to speak in the manner of others, with the second pair, the lesser shoot of the third and fifth pairs. It passes through the foramen of the second pair into the cavity of the eye." *Ibid.*, pp. 234–235: "That called the fourth [inferior alveolar] is larger than the other three and descends through the external surface of the hidden muscle [medial pterygoid]. It reaches the internal foramen of the jaw and, disseminating through this opening, it approaches the roots of the lower teeth. This portion of the nerve slips through the external foramen of the jaw to the muscles, lip and skin lying there. However, it ought to be noted that before the fourth nerve enters the jaw it usually gives rise to a sizeable nervule [mylo-hyoid] which, very long and slender, creeps under the jaw until it inserts into the fourth muscle opening the jaw as well as into those placed beneath the chin, and inserts partly into the hyoid and partly into the tongue. The fifth nerve [lingual], somewhat smaller than the previously described fourth—although sometimes it is found to be larger—crosses over the muscles hidden in the mouth to reach the lower side of the tongue and inserts into exactly half of it as far as the upper surface. But as yet I have been unable to ascertain whether or not the external tunic or crust of the tongue is formed from these nerves. Furthermore, from the same region in which these five nerves arise from the aforementioned third trunk [mandibular division of V], three or four other nervules are extended to be distributed through the muscles hidden in the mouth."

234

NB WH Whether Fallopius is not too curious, pronouncing with certainty that which Nature does in an uncertain way, is done in a certain way.

NB Galen was not discoverer of nerves from brain to senses. Cicero, Tusculanian Disputations, I, p. 339.[845]

Likewise Cicero [has] much [to say] about the use of the parts in De Natura Deorum, book 2.[847]

---

[845] *Ibid.*, p. 228: "When it has arrived here, united with the rest of the nerves, arterioles and venules, at once it separates into two parts. The larger and thicker [frontal n.] creeps through the upper part of the orbit above the second pair between the periosteum and the fat which freely covers the muscles moving the eye and extends as far as the border of the upper eyelid. Here it is divided clearly into two twigs, although this division first began immediately while lying hidden at its origin. The one [supratrochlear] which faces the external angle ascends through a special canal or sometimes a bony foramen into the eyebrow and to the forehead. The divine Vesalius was mistaken in this case since he considered this branch to be a shoot of the lesser branch of the third pair. The other twig [supraorbital], which faces the internal angle, passes out near the aforesaid angle into the muscle closing the eye and must be distributed to the frontal muscles. However, this twig sometimes has a carved channel through which it is carried, surrounded by bone, and sometimes not. Furthermore, the small part of the aforesaid division lying above advances from the foramen of the second pair between the first and second branches of the same second pair, under the muscle which draws the eye upward and under that which raises the eyelid."

[846] Harvey must have been thinking of Book II, which is a discussion of the endurance of pain and of whether pain is a good or evil. The subject is treated philosophically rather than anatomically or physiologically.

[847] The reference here must be to chapters 47 ff. Harvey is here somewhat overenthusiastic in his reference to Cicero.

Index of names and persons
mentioned by Harvey

Adam 72v

Alberti, Salamon 19r, 20r, 27r

Albertus Magnus 63v, 94r

Alexander the Great, 7v, 31r

Andrews, Richard 16r

Aphrodite (Venus) 57v

Appollonius 17v

Aquinas, St. Thomas 94r

Arantius, Aranzi, Julius Caesar 75v, 96v, 97r

Aretaeus of Cappadocia 35v, 84r

Argent, John 33r, 47r, 92v

Argent, John, daughter of 89v

Aristotle 1r, 1v, 2v, 7v, 8r, 10r, 11r, 11v, 13r, 13v, 16r, 17r, 22v, 23r, 26r, 26v, 31r, 33r, 33v, 36v, 37r, 38r, 38v, 40v, 41r, 41v, 42r, 43r, 43v, 44r, 45r, 49r, 50v, 54r, 57v, 58r, 59r, 63v, 64r, 70r, 71r, 72r, 73v, 74v, 75r, 75v, 79v, 80r, 82r, 83r, 84r, 85v, 86r, 86v, 87v, 88v, 89r, 91r, 94v, 95r

Augustine, St., of Hippo 3v

Aurelian, Emperor 31r

Averroes 33r, 43v, 84r

Avicenna 18r, 20v, 36r, 36v, 63v, 81v, 82r, 88v, 91r, 91v, 96v

Bauhin, Caspar 27v, 33r, 33v, 35v, 37r, 39r, 41r, 42r, 45r, 51r, 58r, 68r, 71v, 75r, 76r, 91v, 92r, 92v, 93r, 96r, 96v, 97v, 98r

Bawd, French 48r

Belon, Pierre 45r

Benivieni, Antonio 35r

Benton, Mr. 39v

Bonacurtio, Baron 92v

boy about Holborn bridg 10v

Boy at Padua bitten by dogs 59r

Bracey, John 36r, 39v

Caesar, Julius 14r

Caius, John 36v

Campeggi, Cardinal Alessandro 16v

Cardan, Jerome 3r, 63v

Cary, Robert, Lord 60r

Casserio, Giulio 63r

Cassius 12v

Catulus, Valerius Quintus 85v

Celsus, Aulus Cornelius 60r

Chichester, Arthur, Baron 54v

Chirn, Rowena, Mrs. 17r, 46v

Christ 72v

Cibo, Cardinal Innocenzo 18v

Cicero, Marcus Tullius 84v, 98r

Cicero, Marcus Tullius, son of the orator 31r

Colombo, Realdo 9r, 17v, 18v, 28v, 36r, 39r, 55r, 66v, 68r, 69v, 70v, 72r, 72v, 73v, 74v, 76r, 77r, 78r, 85r, 87v, 91v, 93v, 96r

Courtesan of Padua 29r

Croft, Lady 19r

Crosby, Sir James 21r

David 30r

Diogenes the Cynic 54r

Dionysius the Younger, Tyrant of Syracuse 31r

Drunkerd [of] Hollingbore 31r

Du Laurens, André 26v, 27v, 33r, 33v, 35r, 38r, 41v, 43v, 45r, 52r, 59r, 70v, 72r, 84r, 84v, 89r, 90r, 95v

Dürer, Albrecht 7r

Duretus, Ludovicus 43v

Du Val, Jacques 60v

Erasistratus 35r, 43v, 78v, 91v

Erastus, Thomas 13v

Eustachius, Bartolomeus 45r, 45v, 75v

Fabricius of Acquapendente, Hieronymus 17v, 62r, 66v, 69v, 73v, 95v
Fallopius, Gabriel 3r, 13r, 13v, 27v, 35r, 41v, 42r, 42v, 53v, 66v, 69v, 73v, 87v, 89r, 89v, 90r, 93r, 96v, 97v, 98r
Fernel, Jean 3r, 13v, 21v, 22r, 24v, 34v, 36v, 45v, 46r, 46v, 94v
Fludd, Robert 39r
Friar with ulcerated genitalia 58v
Fuchs, Leonard 3r
Galen 1v, 10r, 11r, 13v, 18r, 18v, 20r, 23r, 28v, 31v, 32v, 33v, 35r, 35v, 36v, 38v, 42r, 43v, 46r, 49v, 54r, 58r, 59v, 61r, 64r, 67r, 68r, 71v, 73v, 74v, 75r, 76v, 78v, 79r, 79v, 81v, 83v, 84r, 84v, 85r, 86r, 87v, 88r, 91r, 91v, 93r, 94r, 94v, 95r, 96r, 98r
German girl who fasted for three years 31v
Gilbert, William 11r
Gillow, Mr. 35r
Gobbo 7v
Gunter, Nan 11v
Gwinne, Matthew 15v
Hardes, Sir Thomas 57r
Harvey, Thomas, father of William 26v
Havers, Mr. 10v
Herophilus 36v, 53r, 90r
Hervey, Lady 63v
he yt eats Bollough livers 31r
Hill, [Mr.] 84v
Hippocrates 3r, 8r, 11r, 18r, 19v, 23r, 33v, 37v, 43v, 68r, 71r, 75r, 75v, 86r, 87v, 93r
Holler, Jacobus 52r, 89r, 92v
Homer 87v
Jasolino, Giulio 40v, 42r, 63r, 72v, 73v, 81v
Johnson, Joan 39v
Julian, Emperor 85v
Laelius a Fonte 46v, 57r
Leoni, Dominicus 75r
long Harry, i.e., Henry Saville 7r
Loyola, St. Ignatius 55r

Macer, C. Licinius 84v
Man behind covent garden 53v
Man cured of rupture 53v
Man in gibbets 44v
Man, old, with cat 30r
Man who counterfeited hysteria 70r
Man with dandruff 12r
Man with gonorrhea 57r
Man with putrid liver 35r
Man with tripe weighing 6 stone 13r
Man with tripe weighing 8 stone 18r
Man with ulcerated testicles 51v
Marius, Caius 85v
Mary pin her cross-cloth 11v
Massa, Nicolò 13r, 15v, 19v, 20r, 22v, 52r, 53v, 58v, 72r
Mercurialis, Hieronymus 93r
Michael Scot 94r
Milo of Crotona 31r
Oribasius 10v
Paduan lacking glans 60r
Patient with wound of chest that blew out candle 68r
Person hanged 71v
Person struck by lighting at Padua 92v
Person with cranium perforated by trepan 89r
Person with double left ureter 47r
Person with ruptured peritoneum 16r
Person with ulcerated bladder 55r
Piccolomini, Aeneas Sylvius 10v
Piccolomini, Archangelo 21r, 26v, 45r, 72v, 73v, 90r, 93r, 94r, 94v, 95v
Plater, Felix 16r, 24v, 31v, 33v. 40r, 47r, 53v, 94r, 95r
Plautus 10v
Rigdon, Sir William 19r, 39v, 70r
Riolan, Jean, the younger 17v, 18v, 20v, 26v, 31v, 40r, 49v, 56r, 57v, 95r, 96r
Rondelet, Guillaume 33v, 58r
Rufus of Ephesus 8r, 58v

Saynts on the knees 10v
Scoller of Cambridg 85v
Shaw, *Mr.* 26v
Shirley, *Sir* Robert 63v
Simpson, J., of Chalis 19r
Socrates 20r
Suetonius 61r

Thyrsites 87v
Torquatus 31r
Trajan, *Emperor* 34v
ij who [suffocated] recently 85v

Ulmus, Franciscus 33v

Valerius 84v
Varolio, Constanzo 57v, 93r, 94r,
  94v, 96v, 97r
Vesalius, Andreas 14r, 28v, 33v,
  34v, 35r, 36r, 41r, 41v, 53r,
  53v, 55v, 56v, 57v, 58v, 69r,
  70v, 72r, 73v, 78r, 87v, 95r,
  96r, 97v
Vill, Robert de 68v
Vitruvius 7r, 16r
Vopiscus, Flavius 31r

Waker, Richard 39r
Wilkinson, Ralph 31r
Woman, dead, with inflammation
  of liver 39v
Woman with blood in abdomen
  25r
Woman with excrement voided
  by uterus 20v
Woman with prolapsed uterus
  55v
Woman with sexual pleasure
  through umbilicus 16r
Wroth, *Sir* Robert 46v

Young, *Mrs.* 36r

Ziska 10v

www.ingramcontent.com/pod-product-compliance
Lightning Source LLC
Chambersburg PA
CBHW030126240326
41458CB00132B/6525